DIGGING
for
DOLLARS

DIGGING for DOLLARS

AMERICAN ARCHAEOLOGY AND THE NEW DEAL

Paul Fagette

University of New Mexico
Albuquerque

Library of Congress Cataloging-in-Publication Data

Fagette, Paul.
Digging for dollars: American archaeology and the New Deal / Paul Fagette.—1st ed.
 p. cm.
Includes bibliographical references.
ISBN 0-8263-1721-9
1. Archaeology and state—United States—History—20th century. 2. New Deal, 1933–1939. 1. Title.
CC101.U6F34   1996
331.13'77'097309043—dc20   95-41727
                                    CIP

© 1996 by the University of New Mexico Press
All rights reserved.
FIRST EDITION

Designed by
Susan M. Walsh

First paperbound printing, 2008

Paperbound ISBN: 978-0-8263-4582-0

13 12 11 10 09 08   1 2 3 4 5 6 7

*This book is dedicated to my late nephew Calvin W. Van Aman.
He will live forever in the memory of those who love him.*

Contents

ix / Acknowledgments

xi / Introduction

xvii / **Prologue**
NINETEENTH- AND TWENTIETH-CENTURY GOVERNMENTAL
SCIENTIFIC INSTITUTIONS

1 / **Chapter 1**
POLITICAL INSTITUTIONS AND PRELUDE TO THE NEW DEAL,
1915–1932

19 / **Chapter 2**
CIVIL WORKS ADMINISTRATION: THE INITIAL RELIEF
INSTITUTIONAL FRAMEWORK

59 / **Chapter 3**
FEDERAL EMERGENCY RELIEF ADMINISTRATION: A YEAR OF
COOPERATIVE TRANSITION

83 / **Chapter 4**
WORKS PROGRESS ADMINISTRATION: A NEW
ORGANIZATIONAL ENVIRONMENT

97 / **Chapter 5**
RELIEF ARCHAEOLOGY: PRACTICE AND
POLITICAL CONSOLIDATION

123 / Conclusion

129 / Epilogue

133 / **Appendix 1**
FERA: PRIVATE AND OTHER MEASURES

139 / **Appendix 2**
FURTHER WPA WORK: INSTITUTIONAL TRANSFORMATION

161 / **Notes**

199 / **Bibliography**

219 / **Index**

Illustrations 71
Map: New Deal Relief Archeology 81

Acknowledgments

The author has drawn upon the talents and the support of many people through the years for this book. My mentors at the University of California, Riverside, patiently guided me through a wonderful graduate experience. My parents and family were a source of never-ending support. My wife, Connie, has always believed in me and for that I am grateful. Others gave strength through friendship. My thanks goes out to you all.

A note of appreciation goes to the staff of the University of New Mexico Press. Editor Durwood Ball has been supportive and cooperative through the entire publishing process. He believed in my work and gave me the chance. Reviewers who gave thoughtful criticism deserve mention as well.

A special acknowledgment goes to Agnese Lindley for her generous financial and spiritual support over the years. This great lady, and Emil Haury, made possible much of what has been accomplished here.

Other financial support was obtained from the University of California Chancellor's Patent Fund and Graduate Humanities grants, as well as from the National Science Foundation.

Introduction

Between 1880 and 1920 the modern American university was integrated into American society as the central training ground for professionals engaged in business, government, and intellectual pursuits. Similarly, the system of higher education increasingly influenced the composition of the federal work force, particularly as the civil service system expanded. Education and expertise, rather than one's political affiliation, became the primary criteria for employment. Concurrently, academics found it necessary to become more politically sensitive if they were to influence government policy, maintain standards, and protect interests.

By the twentieth century, a host of academic disciplines had discovered greater opportunities for research through networking more closely with the U.S. government. Economic, legal/constitutional, banking, and agricultural experts quickly answered the call from Washington. During and after each World War and the Depression, more specialists entered the bureaucratic fold. Throughout and following World War II, to cite one notable example, the physical sciences prospered. Unprecedented sums of money were funneled into military research and development for the Manhattan Project and the Atomic Energy Commission's weapons production programs. In another setting, sociologists and psychologists affirmed their intellectual legitimacy and value, as the country recognized the degenerative effects of racial discrimination, particularly after the *Brown v. Board of Education* decision.[1]

American archaeology, a subdiscipline of anthropology, correspondingly expanded and developed, in large part because of its long-standing relationship with the federal government. This liaison provided both an environment for the practice of this unique science and a political domain within which archaeologists consolidated their professional gains. While growth previously had been slow, American archaeology emerged from the 1930s more organized and professionalized. After World War II the profession boasted many of its own avenues to recognition. Opportunities for publication; greater standardization and cohesion in training practices; more monies for expeditions; national organization for increased interaction, coordination and decision-making; expanded intrastate and regional cooperation; and a much greater political sophistication that led to the creation of political mechanisms were a result of interaction with New Deal relief agencies.

By 1930 American archaeology, compartmentalized both institutionally and interpretively by geographic/culture regions, had existed for almost 100 years. Critical to the field's existence was the relationship between archaeologists in museums, colleges, and universities, and their brethren in the U.S. government. This relationship in the 1930s hastened the evolution of the discipline, prompting to a greater degree than ever before the politicization of American archaeology.[2]

This study investigates the institutional evolution of American archaeology during the 1930s. Its primary focus is the process by which archaeologists created a modern professional science through government, especially at the federal level. Late-nineteenth and early-twentieth-century industrial society underwent significant economic and political integration and centralization. Paralleling those developments was the growth of local, state, and federal governments. American archaeology mirrored these changes, finding pathways to disciplinary integration through its interfaces with government. These precise industrial features provided archaeologists with expanded institutional bases and new interrelationships, including: new channels for communication and cooperation within the community (especially crucial for new classificatory and taxonomic systems, as well as for more broadly based interpretative schemes); heightened national perceptions of the needs of the discipline; and an environment in which a cadre of national leaders could emerge. In order to professionalize in an industrial society, a discipline must adapt to that society's political system. In short, political aptitude is crucial to the maturation of a science, both internally and externally.

This story is divided into two segments. A short introduction places archaeology within both its governmental and academic settings. The second segment, which constitutes the major emphasis of this study, explores archaeology across the entire spectrum of the Depression and New Deal programs. Archaeological relief programs and their surrounding institutional structures provide the framework and means for analysis. The nature of the New Deal experience created a distinct impression of the abilities of both the government and professionals to create and direct policy for archaeology. Thus, how the New Deal occurred was vital to the actions of both in the postwar period.[3]

Archaeology is treated here as a hybrid science, one that combines elements of two different systems. The discipline deals with problems presented by the prehistoric activities of human kind. However, archaeology is also directly linked with the physical and biological sciences, due to the nature of where and how research occurs. For example, methods and procedures utilized in the geologic and botanic sciences may also be utilized by archaeologists in the field for the removal of human artifacts. Accordingly, archaeologists receive professional training in these techniques along with their training in modern anthropological cultural theories.

Archaeologists have through time utilized general sets of rules and procedures to channel investigation and research (a managerial paradigm). But they had failed to explain why human culture evolved in the manner that it did, or how culture acted as the agent of change. For many years archaeologists could best be described as scattered practitioners rather than professionals. A nationally organized scientific society with a university-based center was not present. Interested avocationists in any particular region often comprised one group. College- or academic-based anthropologists who worked in archaeology formed another, government-based anthropologists yet another. In the nineteenth century, access to this specialty was open-ended. No educational standards existed for training in archaeology; literally anyone could be an archeologist. In the twentieth century, academic training slowly came to dominate entrance to the field. The story of the 1930s revolves primarily around the effort by academic professionals to create a unified discipline.[4]

Generally speaking, by the 1930s a comprehensive procedural organizational and training program for archaeology, universally accepted and conducted in anthropology departments, included studies in linguistics; social and cultural anthropology; geology; geography; archaeology field methods; laboratory analysis; survey and exploration of sites; regional and cultural history and time sequences; the use of pottery shards as a benchmark or standard to extrapolate larger regional studies; dating techniques; the creation of institutions to compete for funding; and the subsequent political and communicative organizations that are central to this study. These main components came to dominate in the university realm of archaeology.[5] The unique setting of relief archaeology would allow archaeologists from around the United States to work together and refine these into a common approach.

As it developed in the nineteenth century, archaeology existed primarily in the federal government. But it also abided in small, independent regional and local institutions. Because of the limited communication system and the absence of national societies to prompt such activities, few of these local institutions exchanged ideas or information.[6] What communication that did occur took place at different levels among varied groups. Networks within the academic world provided the key informal mechanism. Ideas, training, and recommendations for students moved through a limited number of people. Government practitioners had their own network. The twentieth century would see a merging of these two. However, local, state-based agencies and regional organizations existed at yet another level. They maintained their own communication arena. Informal contact developed between the groups prior to the 1930s. For example, amateurs would seek advice from the Smithsonian Institution or local universities. Academic archaeologists, though, had little control of formal means of communication. They were required to publish through anthropological journals or eclectic society chronicles. The Smithsonian provided one avenue but generally directed only to its own personnel. Hence, the informal meetings and regional conferences played a major role in bringing practitioners together.[7] The New Deal provided an environment that increased the size and number of such meetings and conferences. More people from all levels and regions then had the means to see and comment on the new work and information. The relief expeditions thus created a larger informal communication venue.

The available number of archaeologists affected the ability of the academic community to foster coalescence in the post-1920 period. The size of the academic community, including both university and government, was small. A survey of the doctorates granted in anthropology between 1895 and 1950 reveals six institutions granting over 75 percent of the degrees (358 out of 476). The questionnaires and communications sent out by the National Research Council State Surveys Committee to archaeological practitioners during the 1920s and early 1930s grew to 300. In the Committee's annual publications, an average of fifty-five institutions from around the country responded. Another related difficulty concerned who could be hired as an archaeologist. In local and state institutions, a wide array of personnel could claim this title. Geologists, sociologists, amateurs, etc., all represented themselves as archaeologists to the State Surveys Committee. In Departments of Anthropology, many personnel conducted archaeological investigations, even when their original training did not emphasize it. Arthur Kelly, director of the Georgia relief expedition in the 1930s, had been trained as a physical anthropologist prior to initiating his efforts for archaeology. Consequently, defined specialty categorizations were not meaningful. Later efforts by archaeologists to influence government policy did not include requests that degreed archaeologists be hired. Instead, they sought anthropologists whose backgrounds included field training and laboratory analysis. This circumstance precluded efforts to determine an accurate number of working professionals throughout the country. Yet, this enabled a small coterie of active leaders to have a dramatic impact.[8] The growth of archaeology in the 1930s was tied to the increased demand for these specialists to work on relief projects.

A final key component in any community embraces leadership. Quality leadership is essential for the advancement of a discipline. Also, a core of personnel who can take supervisory or managerial responsibility and assume higher roles is pivotal. The creation of an available pool with academic training would emerge for archaeology within university, museum, and government circles.[9] The New Deal relief efforts created a unique atmosphere where young and dynamic leaders emerged at a much faster rate.

An additional component is required, however—one that more closely reflects the inner workings of the political/economic society that exists in the United States. A discipline professionalizes as it develops the political means to interface with the various levels of government. Political acumen and organization constitute necessary ingredients to effect cooperation with other academic areas. The capacity to compete effectively for limited funding, much less to gain recognition, is also significant to enhanced professional standing.[10] Archaeologists had to learn how to deal with these types of problems in the relief atmosphere of the New Deal. Successful competition for funding in its political environment became absolutely necessary. The large size of the work programs also encouraged archaeologists to train young graduate students in managing field operations. Both criteria were affected in the 1930s.

Overall, a significant phase of transition in the development of American archaeology took place between 1930 and 1945. This period saw the emergence and growth of a distinctive academic scientific procedure developed and refined in several key universities. Initially, the Great Depression cut funding opportunities needed for continued disciplinary maturation.[11] The discipline then rebounded from the depths of the Depression to reinvigorate itself. New Deal relief labor agencies acted as the means for this rapid expansion.

Archaeology had always existed in institutional frameworks not of its own making. Each through time offered the opportunity for growth, more control of research, and elevated status. With the inclusion of archaeology in Roosevelt's recovery operations, the discipline finally gained a degree of independence from anthropology. New Deal archaeological relief expeditions became the vehicle for greater communication, standardization of field methods, improved laboratory procedures, enhanced sense of community, national organization, and the growth of political organs.

During this entire time span, the Smithsonian never relinquished its influence, even while other archaeologists created the Society for American Archaeology and utilized relief agencies or other mechanisms to address national research problems. The newer generation nevertheless began to match the Smithsonian in power and prestige. With the adoption by other government agencies of a role in archaeological endeavors, a dual institutional complex remained, but with trained experts staffing both. Part of the saga of the 1930s involves the interaction of academically trained archaeologists in museums and colleges with those in federal service. The latter grouping maintained a central position based upon their access to funding, obviously in short supply during the Great Depression.

The premise, then, of this work is that rather than creating the foundations of modern American archaeology, New Deal relief programs hastened the rate at which archaeology became politically and organizationally unified. The greatest legacy of Depression-era archaeology is that it provided fertile ground for the institutional, professional, and interpretive maturation of the discipline.

Prologue

Nineteenth- and Twentieth-Century Governmental Scientific Institutions

In the nineteenth-century archaeology's bases were private societies, museums, some universities, and primarily the federal government. Yet, few institutions conducted archaeological research, focusing instead on ethnologic survey. At the turn of the twentieth century, however, the modern research university system offered new opportunities for anthropology and archaeology. After World War I, academic anthropologists, many trained in archaeology, were employed throughout federal, state, and private museums. At the same time, publishing venues—books, journals, and magazines—expanded. Knowledge and understanding of the pre-Columbian Indian experience grew substantially. Between 1850 and 1920 fundamental transformations occurred that laid the institutional and methodological foundations for archaeology during the 1920s and the Great Depression.

In his deft assessment of nineteenth-century anthropology the historian Curtis Hinsley identified a salient fact. The Smithsonian Institution dominated American anthropology from 1846 until the emergence of university departments after 1900.[1] The content and course of archaeology were tied to the workings of the Smithsonian and, to a lesser degree, the National Museum and the Bureau of American Ethnology. Within this political institutional network, anthropology-associated archaeology manifested several characteristics: reliance on Baconian organizational and empirical schemes; absence of a comprehensive American-focused interpretive framework; and lack of a cadre of trained, professional archaeologists. The fortunes of this federal scientific arm shifted with the prevailing political winds. Forces beyond archaeology's control mediated anthropological and archaeological endeavors.[2]

Anthropological investigation worked within a rigid, hierarchical system. Smithsonian and museum staff members, usually self-educated, pursued research on their own. Membership in this small and self-perpetuating group stemmed from personal contacts or friendships. Because few universities offered instruction in anthropology and archaeology, educational institutions exerted little influence on the early development of archaeology.[3] In the Baconian hierarchy of anthropology a paternalistic elite interpreted data. Others involved in the anthropological enterprise collected the data. Information published by the Smithsonian, and later by government museums, circulated within a select group.

Given the exclusionary nature of this system, private archaeologists and museums (the latter a haven for "outside" naturalists) focused on investigation of extant Indian tribes. The remaining few investigators researched mounds and sites, studied languages, and carried out crude physical analyses that Thomas Kuhn describes as forms of recondite information.[4] Researchers examined data handy to casual observation. Research areas formed from a body of theory were absent.[5] These individuals, none of whom received professional academic training in anthropology, carried on limited ethnological work throughout the country.[6]

Three themes characterized anthropological and archaeological research in this period, whether it originated within or outside the government. First, the lack of a comprehensive concept of North American Indian culture caused artifact and data collecting to become ends in themselves. Museum displays and collections stood as the final product of ethnological and archaeological work. Second, the closed system of investigation in government archaeology encouraged independent inquiry between private individuals and societies. Third, without any broad, interpretive framework for Indian culture, anthropological explanation was usually based on nationalistic rhetoric, on borrowed European classifications (Old World development, e.g., Bronze Age, etc.), or contemporary ideas such as progressive cultural evolution.[7]

Despite their limitations, people who conducted archaeological work in nineteenth-century America identified themselves as scientists and the intellectual fruits of their endeavors as science. However, Thomas Kuhn again cautions calling the resulting literature scientific.[8] They accumulated a body of information sometimes classified by clearly defined and generally accepted criteria. Still, without a cohesive body of theory, individualized, subjective notions drove the practice of archaeology.

The federal government was the foremost practitioner of the "science" described above, with the Smithsonian Institution (SI) its principal arm. Established with the bequest of an Englishman, James Smithson, the SI commenced operations in 1846. The first director, America's leading physicist, Joseph Henry, of Princeton University, labored to bring competent, meticulously conducted science to government service.[9] Given that the SI was only an unofficial federal bureau, Henry guarded closely against sparking controversy and attracting criticism. In more than one instance, he interceded to insure that Smithsonian-sponsored research remained devoid of "unscientific" content.[10] His administration shielded its workers from public scrutiny. Smithson's conception of science developed without interruption; the SI insulated scientists from those least competent to judge them. Yet, this science lacked any genuine internal community. There were no forums for debate and self-criticism. The SI tried to cultivate public respect for a rigid methodology practiced by a learned few.[11] In this closed system, research projects eventuated in public displays and narrative publications. Still, those mechanisms inculcated in the public mind some appreciation of science.

To give form to his notions about science, Henry organized the Smithsonian as a clearinghouse for the dissemination of information. A major undertaking was to open lines of contact between scientists. Most of the research was not anthropological. Any anthropology done tended to be ethnological. A poor cousin to anthropology, archaeology benefited less from Henry's efforts than other research areas.[12] In part, Henry was responsible for this circumstance. In his mind, ethnology, a less significant intellectual pursuit, lacked the scientific credibility of physics, chemistry, or geology. Consequently, archaeology sought refuge in collection, categorization, and museum displays. A dearth of funding produced few trained professionals.

The superficial analysis of data comprised the principal activity of Smithsonian archaeologists. Data and artifacts collected around the world inundated the staff. Developing a method of inventory and creating a classification system emerged as its primary challenge. Organization and sequencing of data became the focal points of attention. Men, and occasional women, labored under difficult conditions. In spite of their conceptual limitations and fiscal disabilities, they provided future archaeologists with a rich source of data on which to lay the foundations of modern anthropology and archaeology.[13]

The Smithsonian was not the only governmental agency to influence the process by which archaeology became institutionalized within the federal structure. By 1847, the Smithsonian Regents pressured for the establishment of a national museum to house and display the important new anthropological treasures. The United States National Museum (USNM) resulted.[14] Once founded, the museum expanded its scope. Under Secretary Henry, museum appropriations increased almost in direct proportion to the march across the continent. After several years of sectional crises in the 1840s and 1850s, which culminated in the Civil War, the National Museum once more renewed its activities. Its structure again revealed the nature of government archaeology and the types of problems faced by its museum personnel in those years. That a myriad of Indian cultures existed became apparent, as artifacts flowed into the storage rooms. Both the Museum and the Smithsonian still continued to focus on categorization.

Given the multiplicity of Native American cultures, ethnologists and archaeologists needed to place the many cultures into a single body of knowledge. Their inability to conceive of Indian culture as distinct from their own resulted in an unnatural compartmentalization of knowledge.

Displays in the National Museum reflected the application of American values to American Indian or foreign cultures and boasted of the perceived superiority of white civilization. The early work of the Smithsonian and Museum demonstrated that in early scientific study, raw data were interpreted through nonscientific ideas. Interpretation was not implicit in the evidence.[15] Archaeology and ethnology stood as handmaidens to a growing American nationalism that by the end of the century fused the industrial wave to the formation of a new social order in the United States.[16]

Prior to 1900, the appointment of Otis T. Mason as the first curator of ethnology in the National Museum signaled one change in the development of science in the United States. His employment not only began a tradition of hiring university-trained personnel in government, it also formalized the association between science and the university. Mason's interpretive viewpoint was only one among many held by those newly trained Ph.D.'s in anthropology in the university systems. His doctorate in natural history shaped his interpretations of human cultures. When Franz Boas visited Mason's collections in 1885, he disagreed with the classification of human culture into biological categories. Nonetheless, science became subject to a peer review beyond government confines. This included the tightly knit band of scientists in the Washington area, including such anthropological luminaries as William H. Holmes, Alex Hrdlicka, and W. H. McGhee. Also counted among its membership was the head of the other major agency for anthropology/archaeology, the Bureau of American Ethnology's leader, John Wesley Powell.[17]

The Bureau of American Ethnology (BAE) was the last leg of the federal institutional triad that sponsored archaeology. The Bureau, initially established in 1879 as a separate government entity, was incorporated into the Smithsonian in the last decade of the century. Its creation demonstrated that the scientific community was beginning to accept the legitimacy of ethnology. Powell was a dominant personality, and like Joseph Henry, he determined the organizational philosophy of the Bureau: paternalistic Baconianism.[18] In the early history of new disciplines, especially those sheltered in bureaucratic enclaves, single personalities loomed large in their formation. Powell was not academically trained. His geological studies of the American West gave his ethnology a unique point of view. Indeed, Powell's ethnology was regarded as an advance in the formulation of anthropological theory. His interpretive modes derived ideas implicitly from the evidence. He focused on cultural achievements that enabled man to adapt to his environment. Such regard to the natural setting further sensitized him to the active intellectual ferment of late nineteenth-century scientific theory largely centered on Darwinism.[19] While contributing to the larger base of knowledge about Native Americans, Powell took the major step of attempting explanation from the data while not applying nonscientific assumptions.

In the new twentieth century the Smithsonian and the Bureau of American Ethnology lost their monopoly of leadership but retained a good deal of prestige. Political pressures and criticisms comprised a partial explanation. By 1900 the overwhelming portion of the Smithsonian's budget came from Congress, not Smithson's bequest. Government science found itself in a tenuous financial position, just as Joseph Henry had feared. Even the clash of personalities played a role here. Proximity to the source of support and power determined the course of the Smithsonian's subsequent growth. With the rise of modern science in universities, furthermore, leaders in all branches of science quickly freed themselves from undue reliance on government if they could do so.[20]

Understood in another way, during the nineteenth century directors in government science failed to provide for institutional continuity. The heirs of Henry and Powell found themselves in a rapidly expanding, complex atmosphere. No longer could they create and manipulate a hierarchical, highly individualized administration of archaeological and anthropological investigation. Nineteenth-century government science, based on intimate, personal relationships within a very restricted community, allowed men of diverse backgrounds and ability to co-mingle but rarely to collaborate. Synthesis of data occurred only at the apex of the structure. Research was based on individual effort, much of it directed by a single person. The Baconian approach thrived in a specialized milieu. When the political and economic environment changed, its failure to adapt was in good part due to its precarious reliance on strong individual leaders.[21] Because of superior organization, universities practiced more consistent and higher quality research than federally commissioned institutions. The Smithsonian lacked internal mechanisms to recruit and train new personnel in charge of ethnology and archaeology, often leaving these tasks to amateurs. Ultimately, the Smithsonian turned to the university for such intellectual nourishment, thereby elevating the latter's status. In spite of its deficiencies, government science developed effective networks. In the geographically spacious United States, any agency or system that centralized science was useful. The increased interface between government and the new universities only aided communication within the sciences.

The legacy of SI archaeology and anthropology was mixed. A distinct emphasis on fieldwork persisted. Henry insisted on empirical investigation and research publication. Archaeologists and anthropologists adopted both of Henry's principals. Although early investigators broached the first tentative classification system, they had much work to accomplish before they developed sound implied scientific interpretation. Powell added the dictum that fieldwork should be a planned and coordinated endeavor. The Smithsonian nevertheless failed to create a cohesive scientific community. In fact, archaeology increasingly institutionalized on state and local levels. National coordination would come in the future.[22]

Between 1880 and 1920 the first attempts to define "scientific" archaeology commenced. Anthropology (and within it archaeology) established itself in a limited number (eight) of colleges and universities. In these years, the field of anthropology was the source for the institutional growth and initial interpretive frameworks evidenced in archaeology. The relationship between archaeology and the federal government changed. Of special significance, academic archaeologists sought to control throughout the country the scavenging of the private "pot-hunters." The government played a key role in defining the nature of archaeology. Congressional legislation granted far-reaching power to colleges and academically trained professionals over archaeological activities in the United States. According to government regulations only professionally trained individuals were permitted to dig. Federal agencies with legal and administrative jurisdiction over archaeological areas hired only professionals. Much of the cultural and physical remains of North American Indians resided upon federal and state lands. Federal regulating of archaeological excavation and hiring of trained archaeologists solidified the relationship between colleges and government.[23]

The universities profoundly affected the development of archaeological research and analysis techniques. Archaeology incorporated a modern managerial paradigm that stressed use of sophisticated investigative techniques and a tightly controlled excavation process. The result was a significant organizational change for archaeology. Newly established fundamental procedures for research and a foundation for training the first generation of American graduate students surfaced.[24]

Archaeology and anthropology benefited from the transformation in several ways. First, both secured permanent institutional support outside the federal government while retaining public funding. Second, academic archaeology escaped the conceptual constraints of museum-based anthropology/archaeology. Indeed, university-supported archaeologists pushed their discipline beyond cataloging and displaying artifacts. Third, for academic archaeologists universities provided expanded opportunities for employment and research as college enrollments climbed well into the twentieth century. Fourth, archaeology and the other anthropological subdisciplines achieved academic respectability. Fifth, with an enlarged institutional base and stronger lines of communication, archaeology more effectively received new ideas. By the 1920s archaeologists had created a core of research principles that regulated the conduct of their discipline.

The new academic environment was more egalitarian and less restrictive than the SI and the National Museum system. The university ideal supplanted the world of the 1800's Baconian hierarchy. Doctoral training, which required that scholars master numerous skills, conduct research, and theorize, profoundly affected the professionalization of archaeology.[25] As professionals took over archaeology, nationalistic notions about race and heredity diminished. The example set by scholars such as Franz Boas and the placement of their students in universities began to change radically the conduct of archaeology in the United States.

Professionals placed considerable emphasis on fieldwork. But they added laboratory analysis, which helped to free them from cultural biases and progressivist assumptions. By 1900, archaeology had become more systematized. Evolutionary Darwinism and scientific geology lent impetus to this process.

European scholars like Boas brought to the United States the latest methods and techniques. American and European archaeologists developed a strong communication network, both formal and informal. Foreign investigators conducting research in the United States spread their ideas and methodologies among American universities. Like the biological and physical sciences, archaeology began to practice team investigation.[26]

Stratigraphic excavation, a product of European geology, was a method adopted by archaeologists. Accepted slowly, the application of stratigraphy to chronological questions was first successful in the research of Alfred Kidder in the Southwest.[27] To interpret accumulating data and create time sequences, archaeologists inaugurated regional meetings.[28] The regional conference encouraged cooperation between archaeologists and the elaboration of broad classification systems.[29]

Archaeology also persisted in a cluster of small institutions. Some of these were museums that were often originally associated with universities. Others were small private groups like the American Ethnological Society.[30] The new generation of university-educated anthropologists eventually took over their leadership. Adherence to the notion of professional training now provided a common link to all of archaeology. The integration of public and private institutions opened additional opportunities for publication and communication. Societies, predominantly eastern, rarely committed themselves to a single avenue of study, instead prompting a wide range of interests. Activities of private societies both sustained and encouraged public interest in archaeology and contributed to the development of communication networks necessary to the academic community. A case in point was the affiliation of the American Museum of Natural History with the Boas-led Department of Anthropology at Columbia University. The association strengthened the link between public and private archaeology and broadened the public base of support for both. Institutions like Harvard's Peabody Museum or the Field Museum in Chicago supplied research and repository facilities for faculty and students alike.[31]

A next significant step forward for archaeology was the creation of the American Anthropological Association (AAA) in 1899. The first national and professional organization in the field, its formation both signaled the emergence of academic anthropology and provided a mechanism for archaeologists to publish their research and cultivate their professional status. Importantly the founding of the AAA involved professionals based in Washington, a fertile field of organizational notions. Members of the U.S. Geological Survey (USGS), the Bureau of American Ethnology, the U.S. National Museum, and the Smithsonian, affiliated anthropologists in the Department of Interior, and an interested educated elite had previously founded the Anthropological Society of Washington. They first proposed the formation of a Society of American Archaeology but, after some thought, decided upon an organization that encompassed a more comprehensive study of man than archaeology alone. Early on, this energetic group published papers in the Anthropological Society. Other publication outlets included the Bureau's publications, *American Antiquarian, Science, American Naturalist, Forum, The Century Magazine*, and *Lippincott's Magazine*. By January 1899, the AAA issued its own journal, the *American Anthropologist*.

A final broadening of anthropology and archaeology's institutional structure was recognition by the American Association for the Advancement of Science (AAAS) in 1917. This organization established a permanent committee on the teaching of anthropology. This discipline had achieved public visibility and intellectual credibility. Approximately forty-five institutions now offered instruction in anthropology and archaeology on a variety of levels, mostly undergraduate.[32]

The stronger academic foundation and larger pool of leadership did not completely end factionalism. In 1919 Franz Boas opposed American intervention in World War I. In "Scientists as Spies," he criticized in *The Nation* the activities of anthropologists and archaeologists engaged as agents for the American government in Latin America. Washington archaeology and museum personalities, including A. V. Kidder, Neil Judd, Jesse Fewkes, and others, opposed Boas. Such animosities fueled interpretive differences and perpetuated divisions within the discipline.[33]

As an important sector of institutionalized archaeology, museums were divorced generally from external strife for two reasons. First, they generated interest and support outside of the academic realm. Disciplines like ornithology had long found refuge in local natural history museums. In archaeology's case, local communities and small colleges created museums to preserve some of their local heritage. That the organizers and contributors to these local institutions were often talented amateurs troubled the professional archaeologists. The idea that any committed or interested person could dig for artifacts persisted for many years to come. In the minds of some practitioners, archaeology was still a museum-oriented endeavor that ought to concentrate on the collection and classification of material.[34]

Considerable public involvement in archaeology produced consequences both positive and negative. Regarding the former, archaeology's greater visibility translated into increased public funding. Membership in local societies and museums, private benefactors, and even local and state governments aided archaeological research and publication. The lists of small societies included almost every town and city in the United States. Of course, these sources of support dwindled during the Depression, although public enthusiasm remained high.

A negative dimension of public support was that it encouraged amateur digs that often caused irreparable damage or destruction to artifacts and sites. Americans traveled to sites and/or read about the Indians of prehistoric America. On-site visits increased with incremental growth in leisure time and the availability of automobiles. Certainly, talented, self-trained amateurs aided the excavation of sites. Yet, many "pot-hunters" dug for profit or mere interest, disturbing sites before proper inspection occurred. Pot-hunting spread with time, causing great concern throughout the archaeological discipline. This problem forced groups to lobby local, state, and federal governments to protect sites. Neil Judd of the Smithsonian called attention to this situation in 1929. Writing in *American Anthropologist*, he commented as the retiring president of the Anthropological Society of Washington,

> Lacking Federal recognition is of national concern. Archaeology in the United States has been, and is still being, exploited by selfish or misinformed persons; it is being fettered by local emotions and further handicapped by obsolete conception as to the fundamental purpose of original field investigators.[35]

According to Judd, federal protection of antiquities alone was an insufficient system. State agencies often measured the value of artifacts in terms of market value or the sheer numbers collected.

Judd believed that information available to the public exacerbated the problem. The public's appetite for archaeological data prompted the release of numerous books and articles by non-professionals. Such uninformed material failed to dispel the amateurish views and misconceptions about archaeology. Especially goading was the idea that digging was a "democratic" activity—anyone could do it! One source of the problem, Judd recognized, was the inadequate organization of archaeology at the academic level. There were simply too few colleges and universities with strong local links locally around the country.³⁶ Also, the only law that did protect some sites, the Antiquities Act of 8 June 1906, applied exclusively to the public domain. Insufficient enforcement and frequent illegal excavations testified to the degree of difficulty if sole reliance on legal approaches dominated. The collection and sale of prehistoric objects became a local industry in the southeastern states. Judd believed that state laws meant to check these actions encouraged this commercialism and vandalism. Writing such legislation were men who had no conception of the nature of Indian culture or how it transcended state boundaries.³⁷

What, then, of the status of archaeology in the early years of the twentieth century? Its institutional segments expanded with an academic sphere as its core for research. An institutionalized systematic method of inquiry continued to evolve. The four elements of government, college, museum, and public began to coalesce. The means of communication available aided these developments. Yet academic archaeology needed to broaden its communication network through conferences, journals, associations, find more consistent and substantial forms of financing, and strive to control more aggressively its destiny. To date this had occurred through a variety of governmental mechanisms, none of which adequately represented archaeology's interests.

Chapter I

POLITICAL INSTITUTIONS AND PRELUDE TO THE NEW DEAL, 1915–1932

The era of the First World War through the early years of the Great Depression profoundly influenced the development of the American industrial system. Within this span, industrial integration, greater economic opportunity, a more cohesive consumer culture, and a host of other modern transformations continued to evolve. American archaeology too found itself swept along into an arena of change. At first came new openings for advancement. Then, as the worsening depression took its toll, a diminished future appeared likely. Within the shifting economic cycles, national communication efforts at first occurred through two means. One venue came through governmental affiliates, specifically the National Research Council. Another operated within larger academic bodies such as the American Council of Learned Societies, where archaeology was classified under anthropology. Subcommittees of such organizations, staffed largely by archaeologists, served to forge a sense of community between academics. Archaeologists from many regions and subject areas worked together for the first time instead of continuing separately in regional or state-oriented groups or under the control of anthropologists. However, this community remained a mixed group. Academic archaeologists trained in anthropology were making headway in appointments in government agencies and universities. They also became more prominent in national organizations such as the AAA. Yet this community was still structured in a three-tier fashion. The local and state components were still dominated by avocational practitioners and staff trained in a variety of areas. The 1920s and early 1930s witnessed attempts to create continuity in action along with the establishment of common goals.

Cooperative movements between institutions and states facilitated information exchange and encouraged wider discussion. Archaeological conferences, for example, the Pecos, became increasingly more important as accepted tools for solving problems and reaching consensus. The meetings brought about the creation of regional cultural sequencing and taxonomic systems. At first spatial and only generally temporal, more specificity emerged through time from additional documentation. Still later, the formulation of interpretive frameworks evolved from culture or state orientations to multicultural and multi-state foci. Use of the conference procedure, with hands-on analysis between peers, suited a small discipline. Material examination by peers evolved from singular interpretation but still superseded measurement against a general theory (paradigm). Contemporary scientific researchers can measure their results against the discipline standard despite scant preliminary interaction with their brethren. Archaeologists, lacking that capacity, used face-to-face assessment by the prevailing experts, thus supplanting a Kuhnian-type medium. Archaeology was evolving through a crucial stage. Having gathered a large cache of data, practitioners sought to derive patterns and chronology. Structuring the data provided common grounds for discussion and training. The first fifty years of the century were spent recognizing and sequencing specific cultural phases. This face-to-face procedure in itself became a part of the archaeologic research process, or its managerial paradigm.

Training in universities and colleges underwent further modification. Graduate education in anthropology emphasized a familiarity with the discipline's broad areas and taught each student to interpret. Exposure to new field and laboratory techniques expanded the archaeologist's skills and increased expectations for field and lab work. As an example, the application of tree-ring dating spread from the Southwest to the East. Growing interest in knowledge from other realms of science such as paleontology resulted in specialist participation in expeditions and more complex management of them by the archaeologist. The increased reliance on specialty skills also led to refinement in laboratory analysis techniques. This in turn made easier a more thorough understanding of collected materials and the dispersal of information.

Graduate schools prepared archaeologists both for teaching careers in higher education and for positions in museums and government. Part of that training still constituted the use of museums as an integral element of fundamental education, as collections were often examined and prepared by students. The growth of such a bond further integrated the two bases.

Relationships between the academic community and the federal government strengthened as well. Employment policies of the National Park Service and the Smithsonian reinforced the presence of academics in government.

Cooperative investigations between universities and these federal agencies also linked the two more closely.

The status of amateurs changed in proportion to the advance of college-trained specialists. State and federal legislation, concomitantly recognizing the primacy of professional archaeology as defined by the broadening institutional paradigm, restricted amateur activities. The public was "educated" through the operation of relief programs and dissemination of published material that described the precise nature of scientific archaeology.

As a consequence of such broad developments, academically trained archaeologists began to utilize a multifaceted professional procedure to organize and control myriad aspects of their labors. Variations naturally existed between those who practiced their craft in museums, government, and education. However, all increasingly came to depend on recognition in a few journals. As such, their actions in the field, at regional conferences, and in training evolved similarly. Differences could be measured by the political ends utilized to serve their home institutions. A strong sense of territoriality and regional identification thus persisted in spite of national organization.

The practice of professional archaeology incorporated similar training regimens: teaching, research, analysis, and relationships involving academic institutions, the public, and the government. This managing paradigm expanded in the 1920–1940 era. It embraced further political knowledge and control of foundation funding by academic archaeologists. Federal and state agencies also demonstrated their awareness of archaeological needs; new laws resulted. This was followed by the entrance of more trained individuals into government service, thus heightening their political influence. A more successful application of the scientific nature of archaeology's distinctive investigative approach ensued, spreading into additional spheres of the community.

The emergent academic paradigm signified more than a rigorous scientific method. As training drew upon other specialties, it became multidisciplinary in dimension; a part of the process of investigation included management of data and sequence of identified cultures. Archaeology moved from use of natural to artificial stratigraphic approaches, with geology, chemistry, biology, physical anthropology, and other areas contributing to the research efforts. Archaeologists came to recognize (as did J. W. Powell) the complexity of the ecological environment and its reciprocal impact on culture. Studies of the environment's effects on cultural change became integral to any study. This resulted in more detailed investigations, drawing again upon specialty knowledge. One specialization, in particular, was photography. Site analysis became much easier.[1] In one sense, then, archaeology's scientific approach became more technologically dependent as more information was gathered in a standardized manner. Consequently, archaeologists were able to exhibit a larger conceptual grasp of the interrelationships between humankind, nature, and time.[2]

On another front, archaeologists slowly began to demonstrate an awareness that such insight stood as a prerequisite for the development of effective national organization. They recognized their position within the larger political spectrum and understood better the need to effect coordination with it. Yaron Ezrahi, a specialist in the sociology of science, phrased it in this way:

> The ability of science to grow and flourish depends no longer merely on the free and successful use of intellectual resources, but also on its adaptability to political action and its capacity to convert its unique resources into effective means of political influence.[3]

Previously, archaeology had failed to create its own specific political mechanism for this purpose. The subdiscipline instead utilized existing structures, the formation of which had excluded or limited archaeologists' input. Therefore, professional identification came only through the parent discipline, with money, opportunity, status, and publications controlled by the larger anthropological community. When academic institutions such as the American Council of Learned Societies recognized the intellectual legitimacy of anthropology, archaeology joined that realm as well. Importantly, it should be recalled that in order for any of this to happen, a strong industrial base with a more centralized, integrated, and communicative culture and government had to emerge. Archaeology as a science evolved relative to the larger American industrial society. The next stage of development involved the National Research Council, the American Anthropological Association, and, later, the federal government Depression relief agencies, all of which facilitated national communication and cooperation. Accompanying this political and organization coalescence would be more standardized approaches to training and research.

The 1920s marked the third stage of the political professionalization of American archaeology. National academic organizations, those affiliated with the federal government, provided archaeologists with mechanisms to further unification. Simultaneously, cooperative research resulted in the conference format. Both means prompted increased communication and discussion. This allowed for the establishment of major time sequences as well as ideas about standardized training. Then, with an increase in university enrollment, more students found jobs, thereby disseminating those specific training methods. Further coalescence of the discipline ensued.

Prior influences continued in this era as well. The emphasis on fieldwork retained its primacy. Yet new methods of analysis and sequencing found incorporation. Schools and societies, dispersed throughout the country, exhibited diverse intellectual interests as strong regional identification persisted. The move toward a national discipline exhibited an asymmetrical pattern. The Northeast and Southwest evolved more quickly than the remainder of the country. Concern over pot-hunting actually increased, generating more legislative efforts for control even at the multi-state level.

Three notably specific patterns characterized this period. First, cooperative research and training endeavors increased while communication networks expanded; more of the major universities pursuing research benefited from new organization structures. Next, classification and taxonomic systems (not borrowed but inferred implicitly) resulted from the research and training. Third, these classified data arrays helped to generate and substantiate standardized language, providing thereby models for similar efforts in other regions. In short, while functioning within academic structures not of its own creation, but with those same entities moving toward a national posture, archaeology utilized the opportunity to expedite its own unification.

The most notable transformation in organized academic archaeology at the national level came about as a result of World War I. Further means of national coordination within the discipline were achieved with the founding of the National Research Council (NRC). By subsidizing research and meetings, the NRC acted as a clearinghouse for archaeologic information, an avenue amplified in importance as available funds declined in the late 1920s and early 1930s.

The Council grew out of the need to coordinate activities of the nation's scientists engaged in the war effort. A separate group, the Naval Consulting Board, previously had attempted to achieve similar goals, but it sought advice only from inventors and engineers. The National Academy of Sciences provided the impetus for the creation of a federal science agency when an active member, the astronomer George Ellery Hale, prompted the establishment of a committee in April 1916. He offered aid to the president in case of war. On 11 May 1918, a permanent Council was founded by executive order. It derived congressional sanction from the National Academy Act of 1863.

When the war ended, the National Research Council incorporated into its membership other academic disciplines. This revised organization derived funds from government and private sources, the latter supplied by the Carnegie and Rockefeller Foundations. The Council now provided scientists of all types with new opportunities for research. Since the peacetime Council focused its attention on already extant societies, archaeology again obtained representation through anthropology.[4] The Division of Anthropology and Psychology of the Council, organized in 1919, dispensed monies for archaeological excavations, mostly in the Southwest. A subcommittee on archaeology, working under the Division, exercised a degree of influence centering around state surveys. An understood archaeological problem revolved around just what sites existed. The survey method became the most widespread preliminary approach to field investigation. Usually such projects failed to take into account the reality of culture distributions that transcended state boundaries. Nevertheless, the publication of the surveys alerted archaeologists about current research across the country and added to the pool of knowledge.

The NRC formalized the survey activity by establishing a Committee on State Archaeological Surveys (CSAS). The duties delegated included: "To encourage and assist the several states in the organization of state archaeological surveys similar to the surveys conducted by the states of Ohio, New York and Wisconsin."[5] By supporting ventures in this manner, a move toward standardizing preparatory archaeologic fieldwork began. Similar surveys appeared in twelve other states within four years. Beginning in 1922 the Committee annually published in the *American Anthropologist* a long series of summaries of the archaeological fieldwork accomplished in North America. A. V. Kidder and Clark Wissler assumed initial leadership, which passed eventually to Carl Guthe.[6]

In connection with the parent National Research Council, the Surveys Committee occasionally took actions partly political in nature. The ongoing concern about preservation and protection of remains precipitated such action. The Committee or state institutions periodically sought to protect threatened materials. This contributed to the growth of political acumen among many archaeologists. The Council itself, in a further aggressive move, created the Committee on Accurate Publicity for Anthropology. It coordinated the actions of anthropologists and archaeologists asked to visit and investigate sites on relatively short notice. The committee, consisting of Roland B. Dixon (Harvard University), Alfred Kroeber (University of California), Leslie Spier

(University of Washington), and Neil Judd (the Smithsonian), quickly became known as the "Minutemen."[7]

The State Surveys Committee also performed the most crucial step in the construction of a communication network for American archaeologists: arrangement of a series of regional conferences. In effect, this systematizing of face-to-face assessment of data further standardized syntax and sequences, an action consistent with the original intent for the Council to act as a clearinghouse of information (much as Joseph Henry had envisaged the original Smithsonian). The origins of this idea trace back to the Central Section of the American Anthropological Association, a body organized in Chicago in the early 1920s. The annual spring meeting of the Association afforded Midwest archaeologists an opportunity to meet and discuss items of mutual interest. Based upon this precedent, a conference on Midwestern archaeology met in St. Louis in 1929. Another followed for southern prehistory in Birmingham in 1932, and a third in Indianapolis in 1935. Similarly, the first Plains Archaeological Conference convened in Vermillion, South Dakota, in 1931. In 1938 the Southeastern Archaeological Conference gathered at the University of Michigan. A fifth Southeastern Conference took place on the campus of Louisiana State University in 1940. A significant portion of the proceedings were published and stimulated further inquiry.

The ultimate value of these affairs, realized in the face-to-face workings of the archaeologists, meant that most of the experts of the day appeared at each session. Archaeology in the 1920s existed as a relatively small subdiscipline. Artifacts, laid out for all to see, found immediate analysis; ideas floated freely. This proximate process continued through the 1930s and 1940s. Individuals from different schools and areas met and discussed pertinent issues. This contrasted starkly with publication outlets, which usually spotlighted a distinct culture area, often within a state. These conferences proved particularly critical to the maturation of archaeology in the South where investigation lagged behind the rest of the country.[8]

Under the aegis of the National Research Council, organizational expansion of archaeology continued elsewhere. The year 1927 marked the founding of a central research laboratory at the University of Michigan. The Ceramic Repository for the Eastern United States, created in the Museum of Anthropology, fulfilled the same functions as the conferences, only on a permanent basis. In academic terms, it represented an archaeological library of shards serving as a valuable training ground for future members of the discipline.

The ceramics donated made possible later cultural cross-comparisons. Three years later, an Ethnobotanical Laboratory formed at Michigan provided an adjunctive means of establishing classification.

Periodic convocation of the conferences and the existence of the Ceramic Repository evoked broader interpretations. Matthew Stirling outlined thirteen recognizable (to him at least) cultural areas in the Southeast at the Birmingham conference in December 1932. At the same time, another group of archaeologists meeting in Chicago discussed the same idea based on a suggestion offered by William McKern in the spring of 1932. As a direct result of these two meetings, a major classificatory step emerged with the Midwestern Taxonomic Method.[9]

This development eventuated from the formation of a self-appointed committee consisting of representatives of the Universities of Michigan, Chicago, and Illinois, and the Milwaukee Public Museum. The original proposal circulated throughout the community, with later modifications in structure and terminology shared at the Indianapolis Conference. Primarily intended to provide a cultural classification followed by a temporal one for use in the Northern Mississippi Valley, archaeologists applied the method to the Northern Plains and the Northeastern area. This action proved possible as more information revealed the wide cross-cultural overlaps that occurred among prehistoric Indians. The taxonomy broke down cultures by utilizing traceable and comparable features. Incorporation of appropriate terminology and gradations of features distinguished materials.

The Taxonomic Method initially appeared applicable for the construction of sequencing of cultures in the Upper Mississippi and Ohio River Valleys. William Webb later used it to classify Kentucky cultures. The system, drawn implicitly from the data, distinguished it from a nationalistic interpretation of the nineteenth-century. The system worked well in the northern half of the eastern region, an area that lacked adequate stratigraphy. In the South, however, the system elicited opposition since geographic/geologic conditions favored stratigraphic analysis. Initial doubts regarding the efficacy of this method appeared, as additional information surfaced during the 1930s and 1940s. As a result, new approaches emerged to deal with the new, broader perspectives.[10]

Factors affecting the expansion of the discipline existed beyond increased interpretive sophistication and the growth of appropriate organizational structures. The discipline remained small in size (generally fewer than fifty institutions were included in State Survey Committee annual assessments), but its effective avenues of communication increased. In such an environment, individuals continued to play formative roles. Carl Guthe, in particular, symbolized the consummate organization man. His great contributions derived from considerable managerial expertise and a full comprehension of the problems inherent to archaeology's design.

Receiving his doctorate in anthropology from Harvard in 1917, Guthe took up residence at the University of Michigan. He guided the Museum of Anthropology, which had been created within the University Museums apparatus, into one of the most active archaeological programs in the country until 1944. By 1928 a Department of Anthropology followed. It trained a legion of students who achieved fame in the postwar period, including James B. Griffin, George I. Quimby, Jr., Albert C. Spaulding, and James A. Ford.

The importance of Guthe to the course of national institution building spread beyond the fact that he made the University of Michigan one of the leading centers in archaeological study. His assumption of the chair of the Committee on State Archaeological Surveys in 1927 gave him the opportunity to travel, study, and ponder solutions to the problems that plagued the discipline nationally. He envisioned a community that would become a truly national one, combining the finest features of museum and academic research.[11]

Carl Guthe manifested his political and managerial talents in still other ways. Along with attempts to assure coordination of research goals and efforts, he understood that the discipline's ability to grow depended on two factors: continued availability of funds and opportunities for investigation beyond the academic environment. Two examples illustrate Guthe's concerns and perspectives.

In 1928 Guthe initiated communication with Secretary Charles Abbot of the Smithsonian regarding a bill pending in Congress. The proposed law provided for cooperation between the Smithsonian and state, educational, and scientific organizations in the United States. Its aim was to continue ethnological research on American Indians.[12] The legislation passed, and President Coolidge signed the measure into law on 10 April 1928. An additional appropriation statute passed on 29 April for $20,000.

The law contained clearly specified limitations. In any year, no state could receive more than $2,000.[13] The Smithsonian, by virtue of control of the disbursement of funds, dictated where research might occur. Administration of these ventures by state agencies served to maintain the sentiment of state identification. Guthe, as the chairperson of a committee expressly intended to foster dissemination of information, offered to coordinate this program.[14] He objected to the Smithsonian's (staffed primarily by anthropologists) exclusive command over the allocation of research resources.

Abbot, in a response to Guthe, shared his ideas on this subject:

> The appropriation for this work is, as you know, far too limited to cover the entire field, and it is therefore the intention of the Institution to allot the funds for projects which promise the most important results. Applications for grants will be made directly to the Institution, where they will be passed upon by a committee designated for this purpose.[15]

Abbot's statement unequivocally indicated that the Smithsonian would not delegate nor share this authority, an act which might have diminished its influence.

Irrespective of the fact that Guthe failed to acquire more power for his committee, the enactment of the legislation proved significant in two other ways. First, the Smithsonian once again maintained its traditional role as the primary arbiter of national anthropological enterprise. After all, where else in the federal system could Congress turn for direction of such an effort? In part, this stemmed from the long-held prestige enjoyed by the Institution. Certain archaeologists understood that in the mind of both the public and Congress the SI functioned as a repository of the finest experts in the country. As such, it represented for many the pinnacle of scientific research.[16] Second, for many archaeologists who contested the legitimacy of this perception, passage of the bill still held positive implications. Funds normally reserved for federal research would now be allocated to state agencies.[17]

Not easily discouraged, Guthe directed his energies elsewhere. He and A. T. Poffenberger, Chair of the Division of Anthropology and Psychology, compiled a study of recent graduates in anthropology, an action spurred by interested members of the Committee on State Archaeological Survey. The survey quantified transformations under way in the discipline for the years 1925–1930, and thus made professionals more aware of the magnitude and direction of such processes. The committee published the results in a memorandum distributed to professional anthropologists and archaeologists.

Most of those polled submitted a response that revealed three significant facts. First, members of the profession expressed interest about recent graduates and their subsequent careers. Second, women now constituted a growing percentage of the student population. They accounted for thirteen of the fifty-seven Ph.D.'s awarded in anthropology. (All archaeologists were trained anthropologists with a specialty. No institution in the country granted doctoral degrees specifically in archaeology). Third, those who emphasized archaeology in their studies still represented a rather small proportion of graduates in anthropology. Men and women accounted for eighteen of the master's and five of the Ph.D.'s given.[18] Of that number, women received five of the master's awards but none of the doctorates. The data also demonstrated the continued dominance of a very small number of schools. Columbia led in degreed student production, followed in order by Chicago, California, Harvard, and Yale.[19]

Aside from Carl Guthe, the activities of A. V. Kidder reflected a further example of how the discipline changed and academic archaeologists slowly gained a modicum of control over research opportunities in a non-government sphere. Kidder attained substantial stature from his work in the Southwest and later in Guatemala. His prominence found reward with an appointment as chair of the Division of Historical Research of the Carnegie Institution. Kidder believed that archaeology deserved equal access to research resources usually dominated by "mainstream" anthropology. This view contrasted with the practices of other private institutions that funneled monies into archaeological activity.[20]

Last, but hardly least, Fay-Cooper Cole likewise was another key transitional figure of the period. After training at Columbia, he joined the Department of Anthropology at Chicago in 1924. He remained there until his retirement in 1947. Dr. Cole's leadership of the department created another nucleus of academic archaeology in the nation. He began his tenure with the avowed purpose of founding a research-oriented department with broad training in graduate work. Cole understood the importance of archaeology (it became his own area of interest after initially studying Southeast Asian anthropology) and provided it with equal status in departmental studies. He introduced dendrochronology studies and ethnobotanical work while emphasizing the use of strong field and laboratory methods. Under his leadership, what became known as the Chicago field method emerged. The process was defined by careful standardized techniques of field survey and exploration, systematic laboratory analysis, and the utilization of the latest advances in other disciplines. The advent of this method, coupled with Cole's ability to obtain high levels of funding, led to growing numbers of trained students dispersing this approach across the country. While much of what was taught at Chicago could be found in other universities, Cole unified virtually all ideas and concepts into a single program. Chicago became a center of learning in the broadest sense. It maintained a high degree of interest in possible cross-disciplinary applications of ideas and tools. During the 1930s, Chicago students became central figures in the establishment of a more cohesive training and research procedure. By the mid-1930s, when relief archaeology leaders needed professionally trained personnel, they consistently contacted Chicago for referrals. These students carried the latest methods and approaches across the country, thus facilitating the movement toward standardization.[21] Archaeology, then, experienced significant growth due to organizational changes, dynamic leadership in colleges and universities, and the ability to utilize government agencies to work on a national level dating from the 1920s. That process came to an abrupt halt with the beginning of the Great Depression.

The economic downturn of 1929 impacted immediately on all levels of the federal government, as well as the states and academia. Governments struggled to meet the needs of the ever-growing number of unemployed as tax revenues declined. Funds for expeditions and publication became scarce. Limited monies went for salaries and teaching equipment.[22] Scant research support often came from private or philanthropic organizations. The Smithsonian also felt the heavy hand of Hoover's cutback approach to government spending. It labored to maintain a minimum level of funding.[23] Archaeologists, like other scientific professionals, had to search elsewhere for support during the first years of the Depression.

During this time an unanticipated experience ultimately facilitated archaeology's later inclusion into the federal relief structure. The origins of that phenomenon resided not in the hub of excitement that grew in Washington, D.C., with Roosevelt's arrival, but down South in Macon, Georgia. There, the rebirth of the idea of relief labor for scientific work occurred.[24] This chronicle provides a case study of the relationship between state societies, archaeology, and the federal government in those early Depression years. The events that transpired helped set in motion an evolving, if complex political relationship that encouraged political maturation in the discipline. Subsequently, greater cohesiveness in archaeologic practice and activity appeared.

The notion of close cooperation between a southern archaeological society and the federal government at first might appear incongruent. A careful examination reveals otherwise. The South, despite common assumptions to the contrary about resistance to progressive educational change, had attempted primary and secondary education reform by the turn of the century. Colleges and universities began to expand slowly, for they lacked adequate financial support. Only a few states had substantial academic institutions. Almost none supported anthropology and archaeology. Consequently, the Smithsonian and northern institutions conducted most of the research in the South.[25]

Local societies, state museums, and amateurs therefore comprised the bulk of the indigenous archaeological infrastructure. As elsewhere in the country, members of societies and museums, generally educated, possessed the economic means to devote time and donate money to this avocation. Their interest stemmed usually from the intellectual stimulation of attendance at college, travel, wide reading, or all three.

Rather unique geographic circumstances also fed fascination with the archaeologic record. The South, like the Midwest and Southwest, was not highly urbanized. It had yet to develop a substantial portion of its land, thus preserving the presence of widespread Indian culture throughout many of the rural sectors after 1900.[26] In these areas, children from all economic levels grew up finding flints and other Indian relics. This circumstance encouraged an ever-present degree of public interest.[27] Contemporaneously, the American public as a whole exhibited a growing interest in its heritage. The attention given to the spectacular finds in Egypt and the preservation of the American colonial past were examples.[28]

Between 1928 and 1933 two historical events of notable significance occurred in Georgia. First, the City of Macon sought to develop relief work. Second, private citizens, primarily in Macon but throughout Georgia as well, labored to protect archaeological relics. These two actions resulted in the rebirth of the idea of using relief labor under federal auspices for archeological investigations.

In the early Depression years, Macon renovated its surface street system in an attempt to supply work relief. The program required the deposit of hundreds of cubic yards of sand and gravel, which the city initially secured from large earthen mounds located outside the city limits. These structures contained within them the remains and artifacts of pre-Columbian Indians. The attempt to protect these sites involved an active, amateur-based, archaeology group.[29]

The Society for Georgia Archaeology drew its membership from Macon and Georgia elites. They included civil engineers, geologists, lawyers, and retired military personnel. Members had grown up in an area rich in Indian tradition and artifacts. They corresponded on an informal basis with the

Smithsonian to seek the latest archaeological information and direction. The group had existed in an *ad hoc* manner for several years before it permanently organized in October 1933.

The impetus for organization began at the National Research Council's southern conference held in Birmingham, Alabama, in 1932. It took final form a year later. This society, representing the first formal institution for archaeology in Georgia, also provided to those concerned about archaeology a political voice. The members maintained influential relations with local, state, and federal politicians.[30]

The five founding members of the Society, Charles Harrold, Walter Harris, and Lint Solomon (all of Macon), Jim Mallory, and Walter Jones of Alabama, met in the course of their different occupations. They found that they shared a common childhood interest in archaeology. Walter Jones was a geologist who eventually became State Geologist of Alabama and Director of the Alabama Museum of Natural History. He worked on Georgia sites and corresponded with the trio from Macon after meeting them at the NRC conferences.

When the group convened, the main topics of discussion usually included ideas for promoting scientific excavations of Georgia sites and plans for the preservation of those relics in a museum based at Macon.[31] The Society envisioned a museum serving both research and public needs. Importantly, this avocational group acknowledged that quality work could only be conducted by trained personnel, suggestive of the elevated level of intellectual legitimacy enjoyed by archaeology. In a critical sense, the Society became an important vehicle for disseminating this concept. Pursuit of their dream led to contact with the Smithsonian on an occasional basis regarding specific problems.[32] This association warmed considerably after the founding of the Society and just prior to the election of Roosevelt. The personal relationships built at this time were manifested in a significant manner by bringing academic archaeology to the South.

The importance of this liaison between the Society and the Smithsonian also took a political form. This organized and interested public group, representing the politically affluent in Georgia, was linked by a common interest in scientific archaeology. Unhesitatingly, they opted to collaborate with the Smithsonian rather than a state organization. In serious financial trouble like other southern states, Georgia lacked extra monies for archaeological endeavors. The Society thus turned to the source most likely to supply funds. The Institution also enjoyed a considerable reputation in the South. Several of the

Smithsonian's resident anthropologists were known to the Macon amateurs, both on a personal basis and through their publications.[33]

These men, but particularly the physician Dr. Harrold, began a long and fruitful correspondence with Matthew W. Stirling, Chief of the Bureau of American Ethnology, and Frank M. Setzler, Acting and later Director of the U.S. National Museum. Actually, the earliest communication dated back to February 1928, when the retired General Harris, now a lawyer, inquired about some specific information on the mounds in Macon believed to be near DeSoto's line of march.[34] The reply by Matthew Stirling resulted in his visit to the area by April of the following year. This established more firmly the relationship.

The members, interested in having the area excavated by professional archaeologists, sought participation by only those individuals who had been trained at recognized universities. They had familiarized themselves with the various publications on Southeastern archaeology and knew much more researched was needed. They therefore solicited advice on just how to achieve this.[35] The Smithsonian again functioned as a clearinghouse for information.

Relief needs and the high costs involved in preserving the mounds fused with other local interests, including the Society's advocacy of a museum. As the Depression worsened and money for any type of work became progressively more scarce, the Macon city government began to quarry the available sites for gravel. The Society successfully petitioned the city to halt this activity. Additionally, the correspondence between the Society and the Smithsonian exhibited greater urgency as the membership stepped up its campaign to investigate the mounds properly.[36]

Another reason they desired outside assistance centered on the unique geographical and cultural conditions present in the South. The region, particularly rich in Indian culture, had sustained long-lived and varied Indian tribes, due to the extensive river systems present. However, the nature of those southern rivers was such that they had been overflowing practically every year for thousands of years. Artifacts and cultural remains tended to be buried under impressive soil deposits and overgrown brush.

This situation stood in direct contrast to the most spectacular Indian remains in the United States, those in the Southwest. Some of these ruins, extant and well preserved, presented archaeologists an easier opportunity to excavate and analyze. But for the South, with more extensive excavation needed, the investment for each expedition required increased manpower and material costs. That burden, combined with the severity of the Depression in the

Deep South, kept archaeological work to a minimum.[37] Again, here the state survey preceded specific site work. Surveys, the usual and preliminary mode of assessment prior to any full-scale dig, revealed only those sites that appeared most promising.

The Society for Georgia Archaeology foresaw such an approach for its state. Dr. Harrold contacted Chief Stirling in November 1932. He asked whether any Bureau work in their area might be possible, since the Society could not attract any local or state funds.[38] Stirling indicated that

> interest of the Bureau in the Southeast is as great as ever although this past year our field work has been curtailed on account of the economy legislation in Congress. We are hoping that our appropriation will be restored before long so that we will again be able to undertake field work.[39]

Harrold would not be dissuaded. He developed a plan for a survey of Georgia. In a letter to John Swanton of the Bureau he proposed to secure information about Indian sites in all of the 159 Georgia counties by using school teachers and pupils, boy scout organizations, and the general public. For this task, the Society intended to rely on standardized forms that could be issued through teachers' groups, the boy scout infrastructure, and the newspapers. Once completed, the forms would be returned to the Society. As he explained it to John Swanton, he wanted the Smithsonian affiliated with this endeavor:

> You probably realize that there is a sort of national fascination and pride associated with the name "Smithsonian Institute" which does not yet apply to the National Museum nor to the words "Bureau of Ethnology."[40]

Matthew Stirling replied on Swanton's behalf about a week later. First, he recognized that a survey would be an important piece of work. However, the Smithsonian would not lend its name to any project outside of its own control; advice could be rendered, but nothing more. As far as reliance on the students and boy scouts, Stirling observed that while it might enhance their education, he deemed it ill advised. Aside from the possibility of personal danger, careless or unscientific excavation would assuredly disturb the sites, thus irreparably damaging the archaeological record. Stirling also warned that information regarding Indian trails and the like was contained in county records and published works. He doubted that many individuals would devote considerable time to such tedious duty. At root, Stirling feared unleashing an uncontrolled public. He cited the example of how the Pennsylvania Archaeological Society attempted a similar task several years before. It resulted in indiscriminate digging and looting. In short, the Smithsonian withheld involvement and recommendation.[41] Hence, although this episode reintroduced to the Smithsonian and other government agencies the idea of using public facilities and labor for archaeological work, nothing of a concrete nature transpired until Franklin Roosevelt took office.

The gentlemen of the Georgia Society and the Smithsonian, awaiting a more opportune time, found it after March 1933. The flurry of legislation and appropriations that made up the Hundred Days gave renewed hope to many, including those who wished to continue archaeological work.[42]

The idea of using other than trained, specialized labor for excavations in Georgia and information about the aborted experiment in Pennsylvania migrated to the National Park Service. The Service generated the original proposal for using Civilian Conservation Corps (CCC) labor or Federal Emergency Relief Administration (FERA) funds in the national parks. In October 1934 Dr. Harrold wrote and asked about Matthew Stirling's familiarity with just that proposal. He offered reference to an earlier Park Service inquiry to the Society about suggestions for such explorations. Harrold, curious about the nature of the work being undertaken by the National Park Service, asked for advice and expressed hope that funds might be available to build a museum to house artifacts unearthed.[43]

From this continuing correspondence, two facts stood out. First, these men stayed their course and retained their faith. Second, the Society still sought the wisdom of the Smithsonian. The National Park Service remained an unknown entity, at least in Macon. Actually, the Park Service had quietly continued to expand its role in government archaeology, a process finalized with Executive Order and supportive legislation in the mid-thirties. However, the Smithsonian was still somewhat guarded in its approach to the NPS. The SI viewed cautiously any other body that threatened its status.[44] Chief Stirling's response to Harrold indicated an awareness of the Park Service plan to use resources from the Conservation Corps or the Relief Administration. He added that the idea to utilize non-professional labor belonged to Ron Finney, Director of the Parks. Two other points came out in the letter. One, the Service would purchase and protect private lands so recommended by professional archaeologists.[45] Secondly, Stirling observed:

It is obvious that unless adequate protection be supplied by the Federal Government, it would be better in more instances to leave the sites in private hands. With this in mind it has been suggested that a committee be appointed which include as one of its members a representative from the Department of Justice who will meet with the idea of revising the Antiquities Act of 1906, so that its present defect of lacking proper means of enforcement will be remedied. This is all in a rather nebulous stage at present, but the National Park Committee is due to present its proposal in a week or so.[46]

Stirling's reply signified that both the Smithsonian and the National Park Service understood full well the necessity of stronger legislation. The Smithsonian even showed a willingness to participate in a cross-agency committee in order to achieve the legal aim, an approach that bore fruit later. But for the moment, the idea that relief labor programs could effect a short-run result (expeditions), and that they could provide jobs by working National Park Service lands, aided in a long range goal—stronger protective legislation.

The next phase of the politicization of archaeology occurred largely through the Smithsonian, and to a lesser degree, the Park Service. The Federal Emergency Relief Administration facilitated the first elements of change, but the Civil Works Administration (CWA) provided the major impetus. Interestingly, the interaction between a small state society and two major federal agencies helped make that possible.[47]

Chapter 2

Civil Works Administration: The Initial Relief Institutional Framework

The pivotal turning point for American archaeology in the twentieth century occurred when professional archaeologists began working with federal relief agencies in the 1930s. The coalescence of the discipline around the standards set by academically trained and based individuals was significantly hastened by their inclusion into the unique atmosphere of the New Deal. As communication increased and more professional people from wide areas of the country came into contact, more cohesive, standardized approaches to research and analysis followed.

Relief projects simultaneously established stronger, broader ties between archaeology and various agencies of the federal government beyond the Smithsonian Institution, and reinforced the presence of the discipline in colleges and universities. The 1930s were, as George Quimby has observed, "The Golden Age of Archaeology."[1] Several mechanisms facilitated this process. While state organizations, museums, and private archaeologic efforts persisted, these efforts became subordinated to those of the ascending college-trained professionals. The unification of archaeology ultimately took form in the creation of the Society for American Archaeology in 1935. The currents of change were reflected in the level of interpretive works that appeared in both the New Deal and after 1945. Unquestionably, the most significant effects achieved came through the direct relationships between Washington and the archaeology community. This occurred by means of the three major relief organizations: the Civil Works Administration (CWA), the Federal Emergency Relief Administration (FERA), and the Works Progress Administration (WPA), along with the Tennessee Valley Authority (TVA).

Certain key matters shaped the following discussion: the organizational structure of the programs; patterns or general guidelines implemented and followed throughout the period; the nature of their evolution; and the impact of these experiences upon the archaeological community. It should be noted that, aside from relief agencies, the Smithsonian again acted as the governmental organization that interacted with the community. While the Institution's role in this relationship changed over time, it remained the central body in the sphere until the mid-1930s. Prior to that time, it judged the scientific worth of each request for a federally subsidized archaeology project and made budgetary recommendations. In this capacity, the SI often functioned as a politically conservative agency primarily concerned with maintaining the centrality of its status and the public image that accompanied it. This political parochialism strained the Smithsonian's relations with other academics. One other point merits comment. Due to the limited size of the professional archaeologic discipline, a small cadre of individuals remained preeminent throughout this period. Recall that the community existed at several levels and in mixed institutional bases. However, practitioners slowly began to achieve unity derived from their common training and through interaction in the colleges and universities. Yet, the process was incomplete. Competing factions, usually regionally defined, persisted.

A logical starting point for discussion of federal relief lies within the Civil Works Administration with its unique attributes. The insight and experience gained from the experiment with this entity constituted a valuable lesson for Harry Hopkins, Roosevelt, and archaeologists. The design and operation of relief labor projects revealed the practicality of using unemployed personnel to achieve a variety of results, including stimulating the economy. The Civil Works Administration represented a probing, conservative stage of the Roosevelt administration, reflecting the philosophy of the First New Deal. The president believed, as did many of his advisors, that the Depression would be brief; only temporary assistance would be required. Roosevelt optimistically attacked the most severe economic catastrophe in American history along relief, recovery, and reform lines, a strategy characteristic of all New Deal legislation.

Congress established the CWA in order to provide immediate aid for the expected harsh winter of 1933–1934. The program, conceived as an alternative to the "dole," was never meant to be permanent. It aimed to maintain the worker's pride and self-esteem by means of a straight exchange: a "fair day's labor" for a wage, a persistent theme throughout the relief era. Thus, the Civil Works Administration operated as an alternative to the direct relief available to some workers through emergency monies granted directly to the states under the Federal Emergency Relief Administration.[2] In both cases, these programs proved more effective in helping the general public than the Reconstruction Finance Corporation begun under Hoover. The latter only hesitantly lent money.[3]

Despite the fact that the Civil Works agency and related programs were intended to function as stopgap measures until the economy began to recover on its own, the Roosevelt administration soon more accurately gauged the severity and the depth of the Depression.[4] The result took the form of the Second New Deal and the institutionalization of work programs under the Works Progress Administration. The WPA design came, in part, from experiences gained from the Civil Works period.

Congress delegated authority to Roosevelt to increase employment quickly under the auspices of the National Industrial Recovery Act (NIRA), 16 June 1933, and simultaneously through the Federal Civil Works Administration. Roosevelt turned to his most trusted aid, Harry Hopkins. He had worked with the Governor in New York.[5] Hopkins was a talented and eccentric social worker with a knack for bureaucratic innovation and honesty. The new law granted Hopkins the leeway "to construct, finance, or aid in the construction or financing of any public works projects included in the civil works program," with an allocation of $400,000,000.[6]

Hopkins' envisioned the Civil Works Administration as a temporary bureaucracy. Thus, he used other government agencies to mobilize the relief operation. With the Veterans Administration placed in charge of disbursement, state and local relief agencies provided lists of the unemployed and selected those judged in need of work. Administrative departments that existed within the government managed and monitored the operations. The Corps of Engineers in the War Department already possessed an infrastructure ready-made for utilization. Hopkins used people already known and trusted from the Emergency Relief Administration. He also appointed field directors responsible only to him. The Smithsonian and the National Park Service, government bureaus with the necessary support personnel and material, had been involved with archaeological endeavors for a considerable time. Hence, the CWA considered their requests for relief support in this realm more favorably in this chaotic period, a crucial point, given that most of the proposals centered on expeditions in the South, where archaeology lacked a well-established university base.[7]

The CWA also established a characteristic trait of the New Deal: strong public relations. The cost of relief was considered huge by the day's standards.

The publicity elements within all the relief agencies made sure that the public was made aware of the great work being performed. The primary consideration was always the number of employed and the practical applications of the work. In the case of archaeology, the scientific worth followed. This practice was carried on by state and local offices throughout each of the major relief agencies. Through these efforts the common citizen was apprised of the importance of professional archaeology.

Harry Hopkins faced the daunting task of putting four million people to work by 15 December 1933. Amazingly, he announced this goal just one month before on 15 November. Given the crisis orientation of the projects, applications could not be inordinately complex. Rapid implementation required simplicity, and directors gave priority to proposals that dictated use of intensive labor. The hallmark of Hopkins' programs throughout the New Deal meant that most of those funds (usually in excess of 90 percent) were spent on wages. This policy contrasted with that of the Public Works Administration (PWA), which usually invested in costly construction projects where a greater percentage of money went for materials.[8]

The Civil Works Administration, at best an *ad hoc* agency, depended on several others. In this realm, Hopkins' management approach mirrored that of Roosevelt. Although he delegated power to subordinate officials and relied on staff help, ultimately he made all major decisions.[9]

Projects received approval based upon a variety of criteria. The emergency nature of the Civil Works program and the power granted to the president to a large extent eliminated congressional involvement in the process. Yet the concept of the relief programs remained a controversial political topic, with fiscal conservatives philosophically opposed to the expenditures. Southern planters anticipated an appreciable diminution of the labor supply. Federal work relief wages vastly exceeded those paid to agricultural workers.[10] Southern politicians, moreover, feared federal intrusion that would upset the carefully constructed hierarchy of home political bases. Such opposition failed to interfere with the project approval process but did influence its implementation. Criticism of these work relief efforts also came from those who supported the concept but balked when originally projected termination dates for short-lived efforts arrived. This situation changed in the more permanent atmosphere of the Works Progress Administration.[11]

When the Smithsonian proposed the only CWA archaeological projects, they concentrated on the South, a region of considerable interest to its staff archaeologists. Site selection centered on consideration of location and weather. Rural areas received preference, as did those with a milder climate. Regarding the former, keep in mind the scant relief available through most programs for the unskilled rural poor. Thus, archaeological plans were considered ideal for diminishing some of this economic distress.[12] In spite of the political dimension of the decision-making rationale, Hopkins wanted to avoid any controversy and wished to distribute money as equally as possible between population elements and states. The fact that most of the work would be carried out by unskilled and semi-skilled labor meshed well with Hopkins' approach to relief: everyone should have the opportunity to work, regardless of their economic standing and skill level.[13]

Hopkins looked favorably on archaeologic projects, for they required short lead times for implementation. An archaeological dig demands a great deal of preparation on the part of the supervisors, but not the laborers. Previous surveys had already established where useful expeditions might occur. The site could operate with little management if the personnel had adequate skills. The main proficiency required for the laborers was a strong back and attention to detail. They mastered the use of a shovel easily. However, careful digging took practice in order not to destroy valuable and irreplaceable evidence. Another factor that made the South a favored location for the Civil Works programs stemmed from the available soil types. Heavy layers of top soil from frequent river overflow made large-scale projects more feasible. This environmental circumstance, combined with the prevalent large mound cultures, meant that the digs required very large pools of labor. Great historic interest regarding the proposed sites bolstered arguments for financing southern archaeological work relief efforts. Coupled with this, local societies could perform some of the necessary advance work.[14] Last, members of the Smithsonian, eager to return to the field, stood ready to mobilize. Two, in particular, Matt Stirling and Frank Setzler, had experience as field directors and possessed familiarity with southeastern archaeology.[15] Therefore, the Smithsonian overcame logistical concerns relative to time and the assembly of a competent staff through further reinforcement of administration by individuals recommended by the Institution. The Smithsonian, in this first relief experiment, exercised a great deal of power, and, importantly, emphasized recruitment of university-trained professionals.

The first phase of New Deal archaeology in the winter of 1933–34 followed three models: the Marksville model, the Smithsonian digs, and Tennessee

Valley Authority archaeology. The Smithsonian Institution first introduced the idea of using unemployed labor on expeditions in Louisiana. Frank Setzler, Assistant Curator of the National Museum, put the concept into practice at the Marksville site.

The 1933–34 archaeological expeditions brought together college-trained specialists, many of them archaeologists. The evolving training and managerial paradigm had varied slightly from campus to campus and among individuals but found itself in 1933 in a more focused, albeit large-scale environment. It had evolved to a much more sophisticated level than ever before. The key to all the methods and training, however, turned on efficient utilization, in such a manner that the site revealed the most information.[16] These data, after all, are what the archaeologist seeks. The importance of removing the data cannot be overstated if any ideas or classifications are to be derived. Carl Guthe described the centrality of the problem in 1935:

> The technique is that of the documentary historian. The site itself is the document; the layers of deposits unintentionally laid down may be considered the lines of the text; while the objects themselves constitute the words. In order to read the story, the archaeologist must carefully preserve the relative positions of the materials in their proper line and with relation to one another upon that line. Just as in a document, if these words are disarranged and lifted from their context the story is lost. The great worry of all archaeologists is that they may fail to see the true significance of the record as they uncover it. As the trowel and the shovel move the earth the record is destroyed forever, and no amount of regret or imagination will ever replace what has been lost.[17]

Excavations consequently began with a carefully laid out survey that divided an area into a grid, with accompanying levels and drawn contours. Field directors determined excavation methods by the nature of the site, for example, layering or sectioning. They also ordered, when appropriate, cleaning and storage, kept careful field records with positions, and maintained data forms that accounted for all flora, fauna, and structural remains, using photography as necessary.[18]

The Smithsonian tested these ideas and their management in a situation unique to the Civil Works period. In addition to the professional responsibilities cited above, the archaeologist had to remain aware of personnel and agency records, instruction, and excavation. Frequent changes of personnel among the laborers caused some disturbances, while fluctuations in the pool of available skilled technicians brought more worry. Laboratory workers and those who worked with skeletons took time to train. Even the lack of standardized CWA data forms brought confusion in field records. In their own unique way, each of the relief sites faced these dilemmas and more.[19]

Individual supervisors overcame complications by developing field and laboratory manuals that trained workers and standardized data, a painfully slow process only partially implemented in the 1930s. The success of that effort can be measured by the emergence of a major interpretation of the archaeology of the Eastern United States in the 1940s. That new conceptual construct was based in large part on data retrieved in the 1930s. However, while records of those sites sponsored by the Smithsonian or, later, by major academic centers were remarkably similar, results achieved from digs under other auspices often did not correlate well with the evolving procedural approaches nor with the classifications that appeared.

Frank Setzler directed excavation of a large site of mounds and a village with funds provided by the Federal Emergency Relief Administration, assisted by James A. Ford. A young man of twenty-two years, Ford already had three years of field experience. Previously, he had worked closely with Henry Collins of the Smithsonian. His and Setzler's venture at the Marksville site provided new information regarding Hopewell-Woodland cultures (defined by distinct pottery and burial methods) that existed in the Upper Mississippi Valley and Ohio River Valley between about 1000 and 1500 A.D. Their success bolstered the position of those who argued for the efficacy of archaeologic relief. After all, from the evidence garnered, the Hopewell culture, usually associated with the Upper River area, was now acknowledged to have spread all the way down the river. Setzler and Ford supervised over 100 men from August to November 1933, proving the feasibility of large scale American archaeology.[20]

Some persons expressed initial concerns about managing so many untrained people. Would lack of training lead to destruction of a site? Yet, conditions existed in the South to make this a successful, if bold, experiment. With a great deal of overlay soil needing removal, heavy machinery could not be used easily because of the potential destruction of the mounds. Negative projections of some individuals appeared not to have analyzed carefully the situation respecting unskilled labor. Absence of well-defined and cultivated skills among laborers did not indicate a lack of ability or dearth of intelligence.

Many agricultural workers had previous experience in methods of planting and harvesting that required skill, care, and sequencing.[21]

Based upon the Setzler and Ford episode, and the stature enjoyed by the Smithsonian, Harry Hopkins approved six similar projects by December 1933. In a letter to Secretary Abbot, Hopkins stated in his usual terse style, "I am happy to tell you that, pursuant to your request, the following projects under your Department have been approved." New Deal archaeology began in earnest.[22] Among the half-dozen sites, exploration of the well-known mounds in Macon, Georgia, was included. Civil Works officials returned approved plans for implementation by 17 December 1933. The sites had to be "dismantled" by March 1934, though.

The original grants, specifying the employment of at least 1,104 persons, expanded to about 1,500. The wages available amounted to $143,000 out of a total fund of $153,000 (or about 93 percent), with the following distribution of labor by state: California—208, Florida—457, Georgia—205, North Carolina—104, and Tennessee—130.[23]

Early communications between Matt Stirling and the Society for Georgia Archaeology indicated how the Smithsonian made decisions and developed policies regarding federal-local archaeological projects during this time. They exchanged concerns about storage of materials and publication of the forthcoming information.

The Smithsonian actively sought the aid of local groups to assure rapid implementation of the short-lived programs. Director Stirling, first in a brief note to Dr. Harrold of the Georgia Society, then followed by a longer letter, related on 7 December 1933, "Approval granted for Civil Works project.... Would cooperate closely with your society stop. Is this agreeable to your group."[24] The communication continued and clarified further the relationship:

> The projects were proposed . . . in five states. . . . All of these proposals of course were contingent upon the necessary permits being obtained.
> . . . Dr. Swanton showed me your letter . . . in which I see that you are already contemplating work on this group.[25]

The Smithsonian had designated Dr. Arthur Kelly, a Harvard graduate in physical anthropology and recently of the University of Illinois Anthropology Department, to direct the Georgia operations with assistance from James Ford of Marksville. Stirling hoped that the work would coordinate well with that already planned. He added:

The Smithsonian was anxious for local support and the Society wished outside help. The field party would be under instructions to cooperate in every possible way with your archaeological group. It is an opportunity to undertake excavation and restoration on a scale that it is not likely to happen again.[26]

Clearly, Stirling could not anticipate opportunities to embark on similar large-scale projects in the near future. From the outset, both men exhibited a mutual eagerness to establish close ties.

Delighted to have such support, the members, particularly Dr. Harrold, energetically secured legal permission from private landowners for access to archeological sites. They also strove to insure an abundant supply of workers. Even the Junior Chamber of Commerce became actively involved, purchasing some Indian property later donated to the city for a park.[27] The other principal concern of the Society, aside from acquiring land access rights, pertained to the disposition of unearthed materials. They had long envisioned a national monument and museum at the Ocmulgee mound sites. Thus, they wanted assurance that the artifacts would remain in Macon. Stirling's favorable response assuaged their fears over loss of Macon's artifacts.

The dialogue between the Society and Stirling exemplified the close working relationship the Smithsonian strove to maintain with officials at the state level. It likewise documented the process by which a local organization sought to extend its influence. The interaction of such institutional components of archaeology formed a critical aspect of the development of the discipline. The usual path trailed through the colleges, but state and local groups played an integral role in the process of national coordination. The relief programs facilitated this process. In turn, the acceptance of the presence of university-trained personnel by state and local bodies reinforced the growing centrality of the college base along with the belief that it should set the standards for "proper investigation." The Macon individuals offered ready acknowledgment of the caliber of the men needed to accomplish their desired goals.[28]

The six projects established a distinctive pattern that proved both advantageous and troublesome for university-trained archaeologists and anthropologists. The main concern focused on the fact that archaeology now would be associated with federal appropriations for relief. Before, it had functioned in a more limited, if pristine scientific setting. Archaeologists later expressed fears in the 1930s that pork-barrel politics would be more determinant in funding research than prescribed scientific and academic methods. This apprehension manifested itself again in the post-1945 period, when legislators had to vote for the Missouri River Basin Survey funding.

The CWA projects saw the erection of two main sites in Florida and one each in North Carolina, Georgia, Tennessee, and California. Stirling described the criteria that applied to site selection in these states:

> The choice of archaeological sites was limited by climatic and economic factors. First, the work must be done in regions where winter conditions were mild, and second, it must be carried on at sites where unemployed labor was abundant and close at hand.[29]

The Smithsonian, staffed by college-trained professionals, chose supervisors who combined previous field experience with university education in anthropology and/or archaeology. Kelly, of Harvard, along with Ford, of Louisiana State University, directed the Macon site. The remaining supervisors comprised Smithsonian staff. Frank H. H. Roberts, an acknowledged field expert of the Southwest, resided in Tennessee. Matt Stirling took charge in Florida with the assistance of his professionally trained brother, Gene. Duncan Strong, Winslow Walker, and a young Waldo Wedel worked in California. W. D. Colburn, Jesse Jennings, and, later, Joffre Coe oversaw work in North Carolina. From Washington, Secretary Wetmore and Frank Setzler voiced the needs of the field personnel to the various federal administrative agencies.[30]

The California excavators worked a site on the edge of Buena Vista Lake, some six miles from the city of Taft, in Kern County. Chosen partially for its favorable weather conditions, Smithsonian investigators had surveyed the lake site some time previously and reported that the field research should be of significant value. A large pool of unemployed men from surrounding oil districts made recruitment easy. With the site accessible by automobile, workers could provide their own transportation. This contrasted with experiences in the South, where laborers often lived at the lowest levels of poverty. Also, southern sites were often remote from main transportation facilities. The Buena Vista area had known almost continuous habitation from 1306 to about 1800. The SI deemed an investigation here pivotal to the classification of sites in the general region. They hoped it might provide further evidence of the Spanish intrusion. They even arranged for access to surrounding areas of lesser but related importance, if needed. Approved to begin immediately, the work had to terminate on 15 February 1934.[31]

An analysis of the original manifest indicates the typical approach inherent in Civil Works archaeology. The job specified 208 workers be employed: 8 supervisors, whose pay would range from $40 to $50 a week, depending on status; 25 skilled laborers; and 175 unskilled workers. Wages made up about 94 percent of the total allotment of $30,105. Other expenses amounted to $1,725. Strong received a denial of his request for a surveyor, a critical function, since no one was available. As a result, he had to hire one engineer to perform two supervisory jobs for a maximum wage of $75 per week, a type of personnel problem that often confronted supervisors.

Still more detailed information about those labor relief efforts that required small capital investments can be gleaned from a survey of the wage differential. Unskilled persons earned $.45 per hour, while skilled individuals received $1.10 per hour. Supervisory pay structure followed this pattern:

a. Assistant archaeologists—2 men at $50 per week
b. Survey and draftsmen—2 men at $45 per week
c. Personnel clerks—2 men at $45 per week
d. Accountant and timekeeper—2 men at $40 per week
1 archaeologist paid by the Smithsonian Institution
1 assistant archaeologist paid by the Smithsonian[32]

Whatever the discrepancies between wages offered skilled and unskilled personnel, one fact remained clear. The project placed an emphasis on the recruitment and use of "bulk" labor required to move dirt.

Since the Smithsonian staff brought much of their own specialized equipment to the digs, the bulk of the materials' cost came from the workers' tool supplies: shovels, picks, trowels, string, pegs, and mattocks (a combination tool with the blades of an ax and a pick), etc. Rental fees were not paid for this land. The great California land combine, Miller & Lux, granted access without charge.

The nature of these expeditions placed great responsibility upon both supervisor and worker. Untrained men often toiled on their own for lengthy periods of time while archaeologists busied themselves with analytical work. The data, according to Strong's reports, indicated that one supervisor could have as few as forty-six to as many as seventy-five men to direct on any given day. Directors offered some prior instruction, yet training derived principally from on-the-job experience. Strong summarized that the work force from which he drew fell into two main classes:

Older men who were not well fitted for heavy road work, and younger men without family responsibilities who had not previously been placed on any relief project. Most of the men had been oil workers in some capacity.[33]

Still others included younger males who had never worked after leaving high school or junior college. Overall, the group represented a large sector of the unemployed labor force during the Depression. The hard times had often destroyed banks and thus the savings accounts of the elderly, the retired, and the older worker. The young, who had no experience, could find neither work nor training. Members of these groups lacked eligibility for any relief, as they had not been in the work force recently. Additionally, individuals in these categories frequently suffered age discrimination. These project had special meaning for them.[34] Strong recognized this and expressed distinctly favorable feelings towards his workers:

> They were a willing, energetic lot, too energetic, in fact, as they were inclined to make the dirt fly, regardless of artifacts. Soon, however, they learned to work cautiously and carefully; and caught the spirit of the mystery and interest of the work. They became enthusiastic over their finds, and developed into a really splendid lot of workmen for an archaeological project.[35]

Generous thoughts indeed that reflected the kinds of relationships that developed during New Deal programs.

On the whole, these words typified the comments reported by the field supervisor. We can empathize with the feelings of men finally able to work and earn money; this alone would have been a small miracle in those Depression times. The men also labored in a totally new environment, one characterized by the excitement of discovery and an element of mystery. This enthusiasm was sustained throughout the New Deal period. The digs themselves, therefore, served a dual function. First, they attested to the accepted importance of professional archaeology for the retrieval of unique data. In addition, they acted as vehicles for the profession's further popularization by exposing countless thousands of Americans to archaeology, an experience they would not soon forget.

The complexity of laboratory work required segregation into specific stages. Trained workers performed one or two defined tasks. Although not instructed as easily, the technicians likewise approached their new roles with vigor. This job classification represented the path by which the middle class participated in the great archaeological experiment of the 1930s. The skills required usually went beyond the capacity of the field worker. As such, Strong drew his assistants, again young men, from Taft and Maricopa. A few came from Berkeley since the immediate area could not supply sufficient laboratory personnel. They observed for a time, then when digging uncovered skeletons, they received practical field experience. In Strong's words:

> They applied themselves with rare fidelity, and spent evenings studying the names of bones in the human body. They developed into real technicians, exercising great care and no little skill in uncovering skeletons and preparing them as specimens. Considering that they had no previous experience in this line of work their results are remarkable.[36]

The size of this expedition provided a unique opportunity for both laboratory and field workers.

Briefly, the two mounds at this site constituted part of a historic Yokuts Village that Spaniards had visited in 1772. The first mound, over 1,000 feet long, 200 feet wide, and 8 feet high, represented a substantial edifice, indicating the amount of labor needed. While the removal of the overlay material took some time, the analysis and detail work demanded even more. Artifacts were sent along to the National Museum for further analysis and preservation.[37] However, the published results of this expedition did not provide any clear direction as to the importance of this site. Waldo Wedel's conclusions indicated that further evidence was necessary before broader interpretations could be drawn.[38]

The southern expeditions, more labor intensive than those in California, also exhibited divergent needs and produced different results. The effort in California helped to furnish information on extensive work previously conducted. California already hosted a strong academic archaeological tradition in which the University of California, the Southwest Museum, and others had long worked the state.[39] The southern states lacked such an infrastructure. But the impetus derived from the Civil Works Administration and other New Deal programs in part ultimately stimulated just that kind of growth.

A project in North Carolina presented an instance where greater and immediate understanding resulted from the CWA efforts. The site included an important Indian mound (215 feet long, 180 feet wide, and 10 feet high) on the Hiwassee River at the mouth of Peachtree Creek, about seven miles above Murphy in Cherokee County. John Swanton of the BAE had visited the site about two years before and thought it could be the town of Guasili. Swanton's interpretation of the records led him to believe that Hernando de Soto had visited the location in 1540 and enjoyed, among other things, a meal of dog meat. William B. Colburn and Jesse Jennings co-directed the dig, aided by an assistant archaeologist. One man fulfilled the task of both surveyor and draftsman, while another functioned as the accountant and timekeeper. The ninety unskilled and ten skilled laborers received the same pay rate as those in California (this parity in wages changed under the Works Progress Administration). Total costs for this project originally amounted to $15,000. Wages consumed $13,800, or about 92 percent, of the initial expenditure. The CWA later allotted an additional $500 for teams of horses and scrapers. Permission to work the mound had been obtained easily through the owners. Again, the materials uncovered ultimately found their way to the National Museum.[40]

A wide array of discovered materials aided in clarifying the importance of this site and resulted in new published analysis. Artifacts included gaming stones, vessels, bone implements, pipes, and shell ornaments. Workers unearthed the remains of wooden structures and stone-lined graves containing skeletons. Of the numerous European objects found in the top layers, none could be linked to de Soto, however. The smoking pipes and potsherds gathered apparently belonged to a type more commonly found in northern Georgia. As a result of the failure to find anticipated types of materials, initial judgment concluded that the site lacked great archaeological significance. Nevertheless, Frank Setzler and Jesse Jennings published the most conclusive report to date on this aspect of North Carolina prehistory. They categorized the two cultures found as Mississippi and Woodland and determined that the Peachtree site had probably been inhabited by the Cherokees from at least 1830. However, the site also exhibited a blending of components so as to make absolute judgment difficult. A revised interpretation of the materials found form in Jennings' doctoral thesis at the University of Chicago.[41] The North Carolina project, then, had the effect of promoting proximately immediate written interpretation.

Keep in mind that single excavations rarely turn up Rosetta stones. Archaeologists, like other investigators, work under the assumption that a great deal of information must be unearthed before any significant conclusions can be reached. The data recovered usually does not lend itself easily to implied analysis, compounded by the fact that it has been subjected to hundreds, even thousands, of years of environmental pressures. Consider, too, that artifacts represent but one aspect of the puzzle. Ideas and analysis of broad areas evolve only after years of collecting. The archaeologists of the post-1945 era drew upon the sources that had been collected in the previous one hundred years.

Archaeologic activities elsewhere in the South also proved important beyond mere research considerations. The Florida excavations further attested to the fact that Smithsonian archaeologists had to become attuned to local political sentiments. Experiences here also demonstrated the importance of public support. Matthew Stirling, a capable field supervisor familiar with the area, directed the two main areas of the Florida operation.[42] Stirling worked from a plethora of headquarters and commanded over 400 unskilled, 45 skilled, and 12 supervisory personnel. His extensive correspondence with the Washington-based Frank Setzler reflected his efforts and the varied responses to the projects, with hostility occasionally directed towards Washington.[43]

The proposed entrance of the federal government into Florida archaeology received more than local notice. In a letter to Assistant Secretary Wetmore, dated 11 December 1933 (before work had begun), T. Van Hyning, the director of the Florida State Museum, inquired about the possibility of some cooperation. He listed his ideas about important sites but dwelt more on the topic of the disposition of artifacts. He opposed their removal from Florida, considering such actions as looting.[44] In response, Wetmore fired off three letters, one to C. B. Treadway of the Florida Civil Works Administration in Tallahassee, to alert the state bureaucracy to the potential problem.[45] The next day he sent a confidential note to Matt Stirling in St. Petersburg urging him to contact Van Hyning. Wetmore also enclosed the original letter from Van Hyning and asked that Stirling proceed with this caution in mind, "He frequently appears obstinate and somewhat unreasonable."[46] In a third letter, the Secretary replied directly to Van Hyning. He attempted to pacify him by describing the proposed excavations and the caliber of the men in charge. He offered assurances that duplicate specimens would stay in the state in properly safeguarded museums, even if the state, then in dire economic straits, lacked the funds to carry out the task.[47] Van Hyning's reply, of a milder nature then his original letter, indicated that he understood full well a fundamental scientific dictum: those with sufficient resources would dig and those without observed.[48] This episode also demonstrated that archaeologists often needed political acumen in such delicate situations to optimize the fruits of research. On another level, circumstances surrounding this series of communications illustrated another theme afloat, one still pertinent to American archaeology. First, they revealed a strong undercurrent of regional identification, confirming the parallel unification of the discipline on two planes. National efforts for standardization in field and laboratory training expanded along with greater communication and new classification and cataloging systems. Simultaneously, problems emanated from the fact that certain individuals jealously guarded their perceived provinces, a direct byproduct of the historic spatial/geographic dimensions of archaeology. The subdiscipline's heritage found it scattered about the country in museums, societies, government, and schools for a century. This fostered strong local and state entrenchment by 1930.

Matthew Stirling remained in St. Petersburg through 28 December, formulating plans for the coming campaign. The management of a large archaeological operation required considerable political and managerial skills not related to excavation. He spent time establishing administrative sources of contact, arranged for recruitment of labor crews, and secured permission to excavate. In one case, he failed to gain clearance but obtained access to equivalent alternate sites.[49] He also worked to coordinate requisite transportation and subsistence requirements, not an easy matter for a project of such scope. Additionally, political turmoil in the West Palm Beach Civil Works office confused the issue.[50] The entire council and director had been dismissed the previous day on orders from Washington. Scandal arose in connection with procedural improprieties (even pork barrel politics exerted an impact). Stirling recounted some of his concerns to Setzler, now back at the National Museum:

> It appears that this Federal project has been considered at Tallahassee as a part of the regular allotment assigned to each county concerned. In short, their own favored projects have to be curtailed to the amount required by these archaeological projects. This does not please the local politicians too much as they feel that it is something foisted on them and something of no visible or permanent benefit to the county, such as the construction of a new bridge, repair of roads, etc.[51]

Stirling wondered if the matter could be taken up with the Washington Works Administration headquarters. He hoped that archaeological projects could be financed separately with an additional allotment. If direct competition for limited funding with other groups could be eliminated, Stirling anticipated greater support for archaeological projects. Truly, this craft worked in a laboratory environment not sheltered within university walls.

The Florida archaeological projects encountered manpower problems as well. Labor assignments came through local offices, where politics assumed an important role; unfortunately, archaeology often received a low priority rating. Hence, the number of laborers available to the supervisors varied from day to day, an irritating circumstance and persistent theme throughout the New Deal. In this realm, Stirling experienced some particularly unique difficulties. Some of the work took place on Canaveral Island in Brevard County, where an insufficient number of laborers lived. Consequently, he sought to recruit men from nearby Cocoa and other locales. However, if these individuals left their towns to work elsewhere, they lost their status on the relief rolls. As a result, Stirling asked Setzler to obtain an exemption, not an uncommon practice, so that such men could work without penalty. He then proceeded to the Bradenton excavations for some actual field work.[52]

Introduction of federal money into the Depression-era South created additional administrative complications. While state personnel presented one type of problem for those in the Civil Works Administration, another thorn in Matthew Stirling's side came from resident spokespersons. He advised Frank Setzler from West Palm Beach on Christmas Day 1933 that local officials were convinced that the programs would be extended beyond the announced cutoff date of 15 February. Moreover, with harvesting continuing at that moment, farmers objected to the fact that Civil Works projects offered higher wages, thus making it impossible for them to obtain sufficient cheap labor. Recall that federal relief wages in this period exhibited no regional or seasonal adjustment. While they oftentimes surpassed but on other occasions failed to measure up to existing wage rates in various parts of the country, one factor remained constant: in the South, CWA wages exceeded the prevailing salary levels. Attempts at conciliation by local Civil Works representatives took the form of cutting back on the quotas of laborers deemed necessary to complete a given archaeological task, a bone to agricultural interests. In addition, these local officials calculated that with smaller available crews, the length of archaeological projects would be extended, thus prolonging the influx of funds.

These actions impacted the Smithsonian immediately. Given the pronounced difficulty in securing sufficient numbers of workers, Stirling inquired about the possibility of structuring an extension. At the same time, he considered retrenching and redesigning the digs to accommodate a smaller allotment of men. The activities of the local federal officials revealed the absence of total, consistently exercised control by Washington. Similar administrative problems faced Hopkins and archaeologists in later work programs.[53]

Frank Setzler, commanding the Washington, D.C., operations, achieved solutions to some of the vexing distractions that plagued Stirling in Florida. He convinced the Civil Works Administration to place the supervisory staff on state rather than local payrolls, thereby enabling Stirling to hire more workers. Setzler, though, failed to gain approval for an extension of the projects, an action that depended on the congressional appropriations process. Simultaneously, he confronted these needs and other tasks while contending with an inundation of unsolicited applications submitted to the Smithsonian. None received consideration. These proposals continued to multiply in proportion to increased media exposure regarding all CWA efforts, including archaeology. Warner Brothers Studios even inquired about the possibility of involving archaeology in a report on Civil Works activities.[54]

Eventually, Director Stirling found time to respond personally to Dr. Van Hyning's original communication. He dismissed what Van Hyning construed as the slighting of state organizations by emphasizing that the need for rapid implementation of the Civil Works Administration policies made expediency the rule: little time existed to distribute the usual formal notifications. He also reiterated Secretary Wetmore's prior declaration that only the most capable men supervised the projects. Stirling then focused on an especially important point when he declared that

> only men who have received higher degrees in anthropology from one of the major universities conducting full courses in the subject are being used for the direct supervisory work in the field.[55]

Consider how far archaeology had come in just two generations. In an earlier time men interested in flint collections secured the attention of a Smithsonian director and found work there. A major step had occurred in the professionalization of the discipline. Increasingly, professionals equated standing in the archaeologic community with solid academic training. The Smithsonian, with Civil Works patronage, fulfilled Joseph Henry's hope that the Institution would function not only as a clearinghouse for information but also serve to further integrate elements of the academic community. The Institution's assimilationist policies brought archaeologists associated with professional university instruction into government relief programs, in a manner beyond mere employment. Dividends also came in a twofold form: data was acquired from important areas not only previously unexplored but in huge amounts. The busy Stirling, in a brief missive requesting more franked envelopes (he corresponded on an extensive scale for a field director!), noted, "It seems peculiar that in our year of depression we are experiencing the most activity on record."[56]

The public popularity associated with the archaeological work had both positive and negative implications. The extent of the publicity generated concerned Frank Setzler. For instance, he felt relieved by the minimal distribution of information about the exceptional material found at the Belle Glade site. Setzler wrote Stirling, "Don't publicise [sic] these results too much."[57] He hoped to avoid both any further controversy surrounding the issue of removal of artifacts across state boundaries as well as additional criticism like that directed by Van Hyning.[58]

Despite the public relations problems encountered early on, it remained clear that in this realm a predominately positive image of New Deal archaeology dominated. Stirling firmly believed that archaeology had elicited more public interest than any other Works' programs. The Florida counties, virtually bankrupt, possessed none of the requisite funds to purchase materials for buildings, bridges, roads, and other improvements. Therefore, men oft times worked at merely hoeing weeds along the highways or the like. Stirling contrasted that activity to the labor expended on the digs, "The archaeological work impresses the local communities as something worthwhile and which enables them to obtain the attention of the rest of the country."[59] Newspapers sent reporters and feature writers from areas beyond the immediate locale of the sites. Stirling commented, "Talks which have been given by the supervisors before local Rotary Clubs and similar organizations have invariably produced an enthusiastic and favorable response."[60] Moreover, a "continual stream" of interested spectators, some of whom had driven hundreds of miles, viewed the projects. American archaeology received the accolades and attention previously directed only to Egypt or the colonial American heritage.[61]

The specific Florida areas investigated included the Belle Glade site near Palm Beach, two sites in Manatee County, and one each in Brevard County, Sarasota County, and Volusia County. The Smithsonian designated the largest site officially as the Okeechobee dig, known more by Belle Glade (its common name), located near Lake Okeechobee about fifteen miles south of Indiantown in Palm Beach County. The Institution originally surveyed the area in 1931 and concluded that its arrangement and symmetry, considered unique in Florida, might provide a possible key to an area little known. As a relief project, Belle Glade entailed extensive camping preparation. The site, removed from the main thoroughfare, dictated that the original transportation of the men and supplies be provided along with a temporary camp, tents, cots, and kitchen equipment.[62]

Gene Stirling, a Harvard-trained anthropologist, managed this sizable expedition. It necessitated more funds and labor than the other sites: two hundred unskilled and twenty-five skilled workers labored along with four supervisors at a fiscal cost in excess of $28,000. Stirling had one assistant, a surveyor and draftsman, an accountant and timekeeper, and a personnel clerk. Wages had been reduced slightly to $.40 and $1.00 per hour, respectively, for unskilled and skilled personnel.[63]

This Everglade habitation, formerly occupied by the Calusa Indians, revealed shards and artifacts of bone, shell, stone, and wood that indicated a static culture. The excavations showed at least six periods of habitation, all evidencing the use of bone implements of arrow and spear points, awls and long pins, deer-horn objects; one site recovered a deer headdress. The proximity to the sea resulted in extensive use of sea shells with ornaments, including pendants, plummets, conch shell hoes, cups, spoons, and a variety of shells, all largely replacing stones. Archaeologists viewed this site as a model of Floridian prehistoric culture. The final report attributed significance to this site due to "the representative collection of habitation shell mound artifacts, burial furniture and skeletal material all from one site."[64]

In Manatee County, Marshall Newman worked Perioco Island and D. L. Reichard directed work on the Little Manatee River. The Palma Sola project, located on Perioco Island between the entrance of Tampa Bay and Sarasota Bay (about 100 yards from Sarasota Bay near Bradenton) stood as one of the most important mounds yet unexcavated in western Florida. The three shell mounds on the island, two habitation mounds, one burial mound, and a small burial area replete with forty-three remains, included one about 900 feet by 120 feet, which explains the funding of $14,000 and the 114 men employed. They found 185 skeletal remains in the mound in tightly fixed positions, well preserved due to their immersion in salt water. Workers unearthed artifacts only in the smaller mound, where animal bones and a kind of sand-tempered pottery rested.[65] Marshall Newman also directed a smaller dig in Sarasota County that housed a mound 110 feet by 13 feet high made of sand. It contained what Matthew Stirling described as "untempered muck and clay with incised and stamped decorations."[66] No stone artifacts marked this site, everything being potsherds "killed" by the boring of small holes.[67] However, the burials and the type of pottery discovered established wider cultural dimensions to this study group.

Excavation of the four mounds in the Little Manatee River site revealed that they ranged in age from the late fifteenth to the middle of the seventeenth century. In the sand-constructed number one mound, built on high ground among scrub pine and other trees, twenty-seven secondary burial skeletons lay. Artifacts included objects of European origin, which dated the mound to post-contact, glass beads, a drilled emerald green pentagonal glass bead, ear ornaments of copper, glazed olive green shards, a conical bangle of sheet gold, and three tubular beads. Mound number two, a circular design of some 65 feet and 6 feet high and slightly predating the first, housed the first cremated burials found in Florida. It consisted of thirty-two cremated and two uncremated children. The final two mounds saw more burials and plentiful amounts of pottery but few artifacts and no European implements.[68]

George Woodbury, assisted by Eric K. Reed and 114 men, focused his efforts on several large mounds near Artesia on the Canaveral Peninsula in Brevard County. Development of this site represented an attempt to add to the slim historical record of these Southern East Coast Indians. An emergency nature surrounded this endeavor: a planned road construction would destroy the Surruque, or Curruque, Indians' habitat by using its shell heaps for regrading. The site claimed a crucial significance because the population of this tribe had gradually decreased after contact with the Spanish, eventually disappearing less than 200 years later (fighting and epidemics were apparently the major contributory causes). Some extant information existed from Spanish records, but the excavation provided more detailed data. Matthew Stirling concluded from the scant information retrieved that the Surruque either had few material possessions or simply did not bury much with their dead. Their diet, more easily determined, encompassed shellfish as the staple, as well as deer, bear, raccoon, and opossum.[69]

Jesse Jennings directed the work in Volusia County for a single mound, about 60 feet in diameter and 6 feet in height, built on a layer of village refuse. They found unique burials on the lower level, arranged in two concentric circles such "that the head on one burial approached within a foot or two of the next."[70] Workers recovered some shards stamped and cord marked but no European artifacts.

Although none of these projects produced a single particularly significant site, as a whole they offered a clearer and broader picture of Florida Indian habitation. Gordon R. Willey later incorporated much of the material from these expeditions into his interpretive work, *Archaeology of the Florida Gulf Coast*. Published in 1949, it constructed the first cultural sequences of Florida Gulf cultures.[71]

The Macon, Georgia, programs formed the focal point of the Civil Works Administration expeditions for a variety of reasons. First, archaeologists lacked adequate information on the Southeast, and Georgia in particular. Incomplete sequencing and a dearth of sufficient, traceable artifacts precluded an understanding of the region. Stirling at one point reflected on the importance of the research:

> More actual work has been conducted on the Macon group than on any in the history of systematic archaeology in the Southeast, and the final results of the work should furnish us with the keys to many of the general southeastern problems.[72]

Second, the projects offered greater rewards than just artifacts and derived information. Recall the activities of the Society for Georgia Archaeology and its efforts to bring an expedition to Macon. This same group focused its efforts on several fronts. One involved the Smithsonian Institution. Another centered upon the development of lines of communication with Georgia congressmen in order to secure a permanent monument and building for the site. A third concentrated on municipal politics as a means of protecting those areas not covered by federal statutes. Incredibly, these remarkable men achieved success in all these areas. Their threefold goals embraced preserving the mounds in their locale and in Georgia generally; assuring that the artifacts remained in the state, as they deemed them an irreplaceable local resource; and constructing a museum in Macon to house the material to attract professional scholars.

The Macon projects indeed became a prominent fixture of sorts. They endured throughout the life of the Civil Works Administration, the Federal Emergency Relief Administration, and the Works Progress Administration, taking final form in the Ocmulgee National Monument. The original allotment of labor reflected the optimistic view held about the Macon mounds: 150 unskilled and 50 skilled (a higher ratio than at other digs, due to the expectation of more cataloging), and 5 supervisory personnel. Arthur Kelly and assistant archaeologist James A. Ford oversaw an expedition funded for nearly $30,000.[73] Kelly's tenure at Macon continued until October 1938, during which time several of the postwar generation leaders, such as Charles Fairbanks, Jesse Jennings, and Gordon Willey, assisted him. Important for archaeology's movement toward standardization of field and laboratory techniques, this site and others brought the country together, figuratively speaking, and allowed common methods and ideas to evolve. Ford, more adept than Kelly in the matter of field training, disseminated the Chicago procedure, which according to James B. Griffin, became the standard approach for southeastern archaeology in this time period.[74]

An illustration of Ford's techniques (for field recovery and artifact analysis) revealed both their importance relative to the needs of southern archaeology and the general archaeologic tactics employed. Overall, James Ford carried three specific strategies to these expeditions: stratigraphy, seriation, and typology. Recall that Ford had trained under the Smithsonian's Henry Collins in the Alaskan Arctic. Collins utilized the metrical stratigraphic method to develop a chronology for that area. Ford brought that to Louisiana and Georgia, where he also introduced a seriation of data combined with stratigraphic.[75] Archaeologists define seriation as the arrangement of data into a time-order series that takes into account the characteristics of the artifact material, whereas stratigraphic analysis deduces only its relative geologic position. Ford used a technique to deal with shallow sites (and thus not amenable to stratigraphic examination) that surveyed the site, followed by surface collecting to determine its chronology. For him, chronology documentation involved the initial identification of the European era, after which one worked backwards. Determination of an implicit classification rather than reliance on the applied logic dominant in the nineteenth century formed the core of Ford's approach.

Trained in the use of ceramics as the primary tool for analyzing eastern archaeology, Ford argued that the change in ceramic styles would be gradual and generally uniform throughout the site survey.[76] He next documented and quantified carefully the differences and arrived at average percentile variations in which he attempted to create a historical model through pottery typology. Later interpretations established regional chronologies on this basis for the first time in the Southeast. As for the pottery itself, Ford divided the ceramic materials into types if the prospect of interpreting time seemed feasible. In his scheme, splitting should not occur unless a spatial or temporal difference existed over features such as paste, temper, methods of manufacture, and decoration.[77]

The Civil Works programs, as well as those that followed, stimulated continuity in the application of such field and laboratory methods. More comprehensive interpretations developed later on as a direct result. James Ford's achievement, in turn, assumed considerable importance, as it aided in the acceptance of the legitimacy of archaeological relief programs. Meanwhile, the professional community watched and waited. Would noteworthy accomplishments at Macon and elsewhere result?

The work at Macon consisted of two specific explorations: one under Kelly at the mounds just outside the city limits, another headed by Ford at the swampy Lamar site about three miles away. Both were located on the east side of the Ocmulgee River. The Macon site had some six mounds, of which four merited additional investigations. The Great Temple Mound (Mound A), one of the highest in the United States, stood 300 feet by 270 feet and about 45 feet high, with a shaft penetrating from the top. Workers entered the second mound (Mound B), the Lesser Temple, 100 feet square, on one side. Mound C, the Funeral Mound, some 230 feet by 100 feet by 25 feet high, housed numerous burials in poor condition, along with some beads and ornamentation. Constructed in five successive levels, with the top sealed with red clay, archaeologists found a skeleton adorned with a disk-shaped, shell-beaded necklace on wooden crosspieces within the bottom level.[78]

Site supervisors regarded Mound D, the Cornfield Mound, 150 feet in diameter, 8 feet high, and divided by a shallow depression, as the most important of the group. Kelly unearthed a circular council house with seats for fifty people in the smaller side, while higher up he encountered a cornfield planted in rows rather than in hills, the technique generally employed by other tribes. Matthew Stirling went on to conclude:

> European objects have been found only in the late village site on the east side of Mound C of the Macon group and in some surface gleanings from Lamar. The De Soto narratives seem to indicate that the Macon and Lamar sites were but sparsely occupied, if at all, in 1540 when the Spanish explorers passed through Georgia. This means the great mounds were wholly pre-Columbian in origin.[79]

Lamar's two mounds and village, on the other hand, suggested "Creek methods of construction." The evidence gathered from the potsherds meant they originated later than the Macon group.[80]

Meanwhile, the Society for Georgia Archaeology pursued arrangements to house the new material. As stated earlier, the Society was a model example of a locally controlled and politically organized group that campaigned for the preservation of archaeological remains. New Deal spending programs permitted the Society to cooperate, first with the Smithsonian, and later the National Park Service, pursuant to these goals. The Society had secured title to the largest mound in the Old Ocmulgee field, Mound A, and the two mounds at Lamar just prior to the commencement of Civil Works digs. The Macon Junior Chamber of Commerce was prompted to purchase them for posterity.[81] After negotiating with the Smithsonian to arrange exploration, the Society planned for the disposition of the artifacts. Members structured a plan for division of the material: half would go to the proposed museum, the other to the "nation's attic." Stirling accepted this arrangement on behalf of the Smithsonian. In the meantime, arrangements to effect storage facilities in the libraries of both the city and Wesleyan College transpired. In sum, the preliminary work of the Society indicated the influence and prestige it enjoyed. Likewise, it demonstrated how a small, determined group could affect the processes of growth and acceptance of archaeology in an area devoid of a major university department.

Dr. Harrold next detailed future plans for more digs in a revealing letter to Matthew Stirling. The Society planned additional excavations in conjunction with the acquisition of city land under the direction of Walter Jones of Alabama and A. V. Henry of Georgia Tech. To complement the expansion of archaeological activity, previously approved development of parks and access roads would occur in the spring. Critical support also came from Harrold's Macon friend, Mrs. Walter D. Lamar, who served on the State Relief Committee. Harrold chronicled her role to Stirling:

> She is very interested in these projects. Most of the mounds are located on the old Lamar plantations (that is they were Lamar plantations for over one hundred years) and in the family of her husband's great uncle.[82]

He carefully added:

> The entire state committee is interested in seeing these projects go through and we will have no trouble securing approval from the state committee for just as much extra labor in time and men as you or Dr. Kelly and Mr. Ford desire.[83]

It should not be surprising that these projects survived beyond the originally specified termination dates, based upon the roles played by these Macon personalities. The political atmosphere in Georgia contrasted sharply with that in Florida. The Junior Chamber of Commerce, for instance, planned within the town to have "pageants, etc.," to help raise money to buy the lands originally acquired with the credit of Dr. Harrold and General Harris.[84] Such evidence indicates the immense local popularity experienced by archaeology, a factor that contributed to its growth in the 1930s.

Elsewhere, the state Civil Works machinery directed further efforts to assure continuance of the Macon work. Gay Shepperson, a senior director for the Georgia State Civil Works office, proved particularly helpful. An articulate, educated women, she enthusiastically supported scientific projects like those at Macon. Her support, it has been suggested, stemmed from her college background. Almost alone, she recognized the need for professionals, including those in the sciences, to work and receive support. Other officials in similar positions limited their efforts to securing assistance for laborers and other workers.[85]

The influence cultivated by Harrold and his organization coincided with the growing interest of the National Park Service to preserve and protect historic and prehistoric sites around the country. By February 1934 Harrold actively strove to coordinate the two movements. Georgia representative Carl Vinson, now a close acquaintance of the Society, introduced legislation that same month to establish a national park at Ocmulgee. Later passage ultimately created the Ocmulgee National Monument.

Interaction among these various federal and state agencies mirrored the developing political strength of archaeology. A support network had spread beyond the confines of the Smithsonian and the Park Service. National visibility of archaeology, associated with relief, had increased, but a large question loomed: what would happen when the relief ended? The accomplishments of the Society for Georgia Archeology attested to the strength of local and state-based groups, at least in the South. It should be noted, moreover, that the major academic organ for archaeology, the American Anthropological Association (the Society for American Archaeology was not founded until 1935), did not coordinate any of these actions, nor did the National Research Council.[86]

Meanwhile, Civil Works projects continued in Tennessee when Frank H. H. Roberts, Jr. and his assistant, Moreau B. Chambers, initiated the first phase of the most extensive archaeological work in the New Deal era farther up in the Tennessee River Valley.[87] The site exploration within the Shiloh

National Park constituted a prelude to the later huge explorations undertaken by the Tennessee Valley Authority. Interest in the Shiloh area dated back more than a year to when the Tennessee State Archaeologist, P. E. Cox, inquired of the Smithsonian if any monies and/or staff could assist him in an inspection of this locale. The Smithsonian seemed a logical choice to consult since it dealt with other federal agencies connected with this region. Cox, in the meantime, worked through the appropriate divisions of the National Research Council while he simultaneously attempted to influence legislators in hopes they might pass a bill that would appropriate money for further work. In addition, the indefatigable Dr. Guthe contacted different organizations that might support the legislation.[88]

Once again, archaeology's spokesperson exhibited increasing awareness about sources of potential support. Cox, working from the state level, attempted political persuasion. On the surface, it appeared that the Committee for State Surveys, the NRC, and the Smithsonian strove toward a common goal: continued archaeological investigation accompanied by support from the state tier. However, a deeper significance drew from Cox, and more so Guthe, who were marshaling support to work on their own as well. This action challenged the autonomy of the Smithsonian in a rearguard action. Remember, the Abbot–Guthe confrontation in 1928, an indication to university-based archaeologists that the Institution's actions reflected its own self-interest.

The Shiloh site, located near Pittsburgh landing on a high bluff above the west bank of the Tennessee River, contained largely unexplored mounds. The artifacts collected went to the National Museum, for this was federal land. The $21,000 successfully solicited by the Smithsonian covered the expenses for 100 unskilled and 25 skilled workers, 2 assistant archaeologists, and the usual support crew. The original grant specified cooperation with Park Service officials over matters like non-disturbance of military burials.

The most prominent features of the site included seven mounds, six habitations, a burial, and a protective palisade with bastions or watch towers. Trenching revealed some thirty houses, a temple (oval in shape), and several refuse deposits. Houses averaged 16 feet in diameter with walls 8 feet tall.[89]

Skeletal remains and artifacts uncovered in the burial mound disclosed thirty skeletons in a flexed position. Specimens recovered included large quantities of shards, stone knives, blades, drills, spearheads and arrowheads, bone implements, shell ornaments, and a host of Civil War artifacts, but no copper. Pottery exhibited a division between shell-tempered and an older grit-tempered grouping, with some cord marked and a larger amount smooth-surfaced. The area apparently served both as a refuge for village sites and a ceremonial center because of its high position. This accounts, in part, for the wide variety of types of cultural artifacts associated here: Fort Ancient Culture in Ohio, Arkansas type red ware, "practically all common forms of Mississippi Valley implements," and Cherokee as well.[90]

Despite the nature of these findings, Setzler felt that if any of the Civil Works sites should be discontinued, Shiloh would be the one. He wrote in January 1934:

> Up to the present time Frank [Roberts] has devoted most of his excavations to village sites around the mounds. In these excavations he has had difficulty because the forester will not allow the removal of any trees. Certain photographs . . . showed a veritable forest on as well as around the mounds. It would seem very useless to try excavations in the mounds if he is not permitted to remove the trees.[91]

Regardless of this complication, in the space of a little over two months Roberts gained a great deal of information on several key sites.

The Civil Works Administration had accomplished many of its goals in a short duration. The projects had set in motion a current of change, particularly in the case of Macon, that continued for some time. The expeditions garnered a high degree of publicity, often national in scope. However, none of the other sites discussed gained the attention directed to the most spectacular expeditions of the decade: the salvage (emergency) archaeology of the Tennessee River Valley.

The Tennessee River expeditions, conducted under a separate administrative entity from the other relief agencies, entailed different management boundaries. The Tennessee Valley Authority (TVA), responsible for the regeneration of a large geographic region, encompassed economic, political, and social spectrums. Transcending state lines, workers carried out salvage archaeological projects under severe time strictures since dam construction followed closely behind. Even money for equipment and wages came under separate budget guidelines.

The immensity and the multistate nature of these operations created a unique environment never before experienced in American archaeology. Several universities and their personnel were involved. Each possessed specific responsibilities for laboratory or museum work, thereby bringing archaeologists together in a way that no conference could. Men worked together under new conditions requiring reliance on novel approaches. Ideas disseminated much more rapidly since supervisory personnel from all over the country arrived to work in the South. While the Tennessee Valley operations dominated the relief archaeology of several states, southern archaeology as a whole experienced significant maturation. Even with designated states carrying out assigned tasks, the Authority programs helped to create a stronger regional identity, one that extended beyond state boundaries. In short, the TVA conducted salvage operations under conditions that stood apart from those affecting the other relief work efforts; these projects must be viewed, then, in a slightly different context. To cite one example, time constraints often precluded a more thorough and time-consuming examination of archaeological sites.

The Tennessee salvage investigations underwent three phases of operation. The Civil Works Administration provided the workers for the initial survey, then the Federal Emergency Relief Administration funded the labor force for the first recovery projects. The Works Progress Administration then sustained that role into the early 1940s. Grouped together, these agencies linked administratively the Tennessee works over the ten years. The excavations opened up a whole new chapter in the prehistory of eastern America. The river valley, an area rich in Indian habitation, had undergone little investigation prior to the 1930s. The TVA represented an administrative advancement beyond the level occupied by the Works Progress Administration or the CWA. That is, the WPA or the CWA were state-based, while the Authority worked as a truly regional conception, both politically and archaeologically.

President Franklin D. Roosevelt asked the Congress to create the Tennessee Valley Authority on 10 April 1933. The Authority organized a multipurpose program to build dams for reservoirs in order to control floods and generate cheap, abundant hydroelectric power. The concept amounted to more than simply a dam-building, fertilizer-producing endeavor; it stood as a bold experiment in social planning. A federal agency committed itself to the total redevelopment of a major region, in part based upon earlier congressional proposals rejected by Republican presidents in the 1920s. In the First New Deal, the legislation passed the House by a substantial margin. In the Senate, its paternal advocate, George Norris of Nebraska, guided it to approval. Roosevelt signed it into law on 18 May 1933.[92]

The fact that archaeology elicited any attention was not surprising, given the comprehensive conceptual approach inherent to this project. Historian William E. Leuchtenberg described the breadth of the Authority as the most spectacularly successful of the New Deal agencies, not only because of its achievement in power and flood control, but because of its pioneering in areas from malaria control to library bookmobiles, from recreational lakes to architectural design.[93]

Salvage archaeology became part of a dynamic, corporate government structure that encouraged innovation.

Archaeologists familiar with this area understood that access to many prehistoric sites would be lost forever. Approved plans called for the inundation of several areas and the building of dams along the river basin, all 41,000 square miles of it. Exchanges of information soon commenced between lower-level personnel of the Authority and the Bureau of American Ethnology. The Universities of Tennessee and Kentucky inquired about the disposition of the sites, while the ever-present National Research Council spokesperson Carl Guthe acted.[94] Accordingly, A. T. Poffenberger, the chairman of the Division of Anthropology and Psychology, wrote to E. S. Draper, the Director of Land Planning and Housing for the Authority, to arrange a meeting with the Smithsonian and the National Research Council to work out a possible solution to the problem of the flooding of the sites. Draper, having shown a prior willingness to negotiate, had also demonstrated an understanding of the problem.[95]

This activity signified other important changes, as Guthe, Stirling (representing the Smithsonian), and Poffenberger assumed responsibility for determining the needs of the entire discipline. Occupying important positions of leadership, they seized the moment. True, they formed but an *ad hoc* association. Nonetheless, it constituted a great leap forward, for their vision extended beyond local problems. This unique action called forth a spirit of cooperation, made possible by the limited size of the archaeological community, the close relationships formed by its leadership, and ambiguity about the new relationship with the Authority, and guided by the generally optimistic spirit shared by the New Deal proponents. While this particular composition of personnel shifted periodically, it still marked a momentous change. By the end of 1933, two trends dominated: a growing usurpation of the once almost solitary power of the Smithsonian in the federal archaeologic realm; and the growing ability of archaeologists to take advantage of shifting political situations. Consequently, the influence of university-based archaeologists grew, as they struggled to codify the aims of the profession.

In another sphere, previous actions had already occurred, with the Smithsonian Institution again occupying a central position. Here, Stirling and Guthe attempted to influence policy through their Authority contact, Draper. Matthew Stirling received a commitment from Draper in early August 1933 to

> issue instructions to all his engineers and field men to take such reports and to give such information as you may request, and to send such information to Science Service or to local minute men or any other.[96]

Such informal communications eventually resulted in the creation within the National Research Council of the Sub-committee on the Archaeology of the Tennessee Valley. A subgroup of the State Archaeological Surveys body, it functioned outside the Smithsonian's direct control. Matthew Stirling of the Bureau of American Ethnology chaired it. Neil Judd and Burnham Colburn assisted. They offered a preliminary plan for a survey of the sites prior to the initiation of any actual excavation. That proved impossible, however, because of the speed with which the Authority began work. Instead, they convened a conference in Knoxville to assess the problem.

Neil Judd represented the Smithsonian, Burnham Colburn and Harcourt Morgan the TVA and the National Research Council, respectively, and they were joined by other personnel from the Universities of Tennessee and Alabama. A new plan came out of this meeting. Judd outlined a proposal that called for an excavation of representative sites based on a Council survey of Wheeler Basin, one made the previous summer by Walter Jones of Alabama. The specter of imminent flooding by the new Wheeler and Norris Dams mandated that the areas be excavated as quickly as possible, utilizing Civil Works Administration labor and tools. Conference participants recommended that a similar survey be conducted immediately at the Powell and Clinch Rivers behind the Norris Dam, accompanied by a complete photographic record. Adequate leadership by a qualified archaeologist and field director with "graduate trained" assistants held the key to the success of any such efforts. Judd concluded that the logical repository for the expected huge collection of artifacts would be the University of Tennessee, with all students of the region permitted access to them. He went on to recommend W. C. McKern of the Milwaukee Public Museum for the position of director. McKern, one of the formulators of the earlier taxonomic system, declined. The Authority then offered the position to William S. Webb of the University of Kentucky. The entry of Webb into New Deal archaeology ultimately established him as one of the more important, central figures for archaeology in this Golden Age.[97]

William Webb possessed several characteristics well suited to his appointed position, ones that also benefited American archaeology. He had highly placed political contacts, tremendous leadership ability, a constant drive to succeed, a love of inquiry, and an extremely stubborn nature; reliance on the latter trait often neutralized those who impeded his "mission." In addition, he entered the field of archaeology through alternate channels. Trained as a physicist at the University of Kentucky, at age twenty he graduated with a master's degree. He then attached himself to the Government Land Office in Indian Territory since he was too young to teach. He remained in that position from 1904 to 1907, rising to the level of a deputy U.S. Marshall. During this period, he developed a passion for Indian lore and history while learning much of Indian life. Eventually, he became fluent in Seminole. By 1907 he returned to Lexington and began teaching in the Physics Department. He continued in that capacity, except for a short further stint of graduate training in physics, then eventually became a full professor in 1913. Next, he enlisted for World War I and stayed on the home front. He rose to the rank of major, his title for most of his life. He retired as the assistant director of the School for Fire Artillery at Fort Sill, Oklahoma. Upon his return to Lexington, Webb continued a close working relationship with Arthur Miller of the Geology Department, one established before the Great War. The two devoted many hours to digging in the hills of Kentucky: Webb was fascinated by the artifacts found, while Miller looked for rocks. Soon William D. Funkhouser, then Dean of the Graduate School and head of the Department of Zoology, joined Webb and Miller. Webb and Funkhouser shared an avid interest in Kentucky archaeology. Each summer they organized an expedition, which included some adventurous graduate students. By 1925 both Webb and Funkhouser had published reports of their work. A net result of these independent field expeditions was that the University of Kentucky wished to receive National Research Council grant money for state survey research. That prompted creation of the Department of Anthropology and Archaeology on 27 July 1927. Funkhouser and Webb received appointed professorships in the department, without pay, but retained paid positions in their respective departments. Webb ultimately became the leading archaeological figure in the state.[98]

The Major effected a rigid systematic approach to archaeological problem solving. The hallmark of Webb expeditions was the proximate military manner in which he laid out his campaign, utilizing the tools of precise organization and designation of clear-cut areas of responsibility. Webb, like Boas, emphasized the laboratory, or in this case, the field. His organizational talents allowed him, aided by a corps of carefully chosen supervisors, to direct huge expeditions. An astute judge of men, he recognized talent and encouraged it. In reciprocation, those assistants became loyal allies. He proceeded to attain a position of leadership within the archaeological community through his Tennessee River Valley work. He likewise represented his new discipline ably at the national level.[99]

After he assumed leadership on 6 January 1934, work began on a survey of Norris Basin, with the first area flooded on 8 January. A survey started in late December 1933 was already in progress, supervised by Burnam Colburn of the Authority. Thomas M. N. Lewis, a talented amateur, conducted the new survey, based upon his background in Tennessee archaeology. Webb's field assistants included Robert Goslin, William G. Haag, H. M. Sullivan, A. P. Taylor, Wendell C. Walker, Charles D. Wilder, George D. Barnes, and A. E. Wilkie; the large number of assistants testify to the dimensions of the project. The other Civil Works archaeologic projects employed significantly fewer supervisory staff. These men worked with a Civil Works crew, often under difficult winter conditions characterized by snow, zero weather, or excessive rains and flooding. The hiring of these assistants followed a pattern present in the other projects. Or, as Carl Guthe wrote, "Young College men trained in archaeology and having experience in field work. They were drawn from university and museum work to meet this emergency."[100]

Salvage archaeology occurred under the most pressing time constraints. Webb readied himself to tackle the task at hand. Since a preliminary survey had been completed, the Major anxiously awaited the initiation of the actual excavations. However, the Civil Works Administration ceased operations in the spring of 1934, causing the dispersal of vast numbers of workers. Webb faced an immediate labor shortage. His response entailed the development of an acceptable proposal to have the project supported by the still existent Federal Emergency Relief Administration. He maintained a pared-down staff of his supervisors, coupled with a crew made up solely of full-time workers; he hired only Works Administration laborers and for limited periods.[101] The completion of the survey itself, then, marked the end of the Civil Works Administration archaeological work relief experiment. The stage was now set for the next phase of this endeavor, under the FERA.

The Smithsonian Institution and the Tennessee Valley Authority exclusively carried out Civil Works Administration relief archaeology. The directors, already on staff, proselytized current investigative ideas. Their younger assistants, most of whom had just left or entered college, adopted similar approaches. Practice in the field further reinforced training. Staff observed the standardized practices of the Smithsonian personnel, practices also derived from academic preparation. Matthew Stirling had attended Berkeley, while his younger brother received his education at Harvard, to cite one example. These early projects allowed the expansion of academic archaeologic standards with regard to training and application of field and laboratory methodology. They spurred growth of political and organizational skills to other parts of the country as well.

For virtually all of these men, fieldwork marked the crucial first step in the analytical process. The second phase consisted of laboratory examination. By this time, an archaeologist faced a whole series of tasks to perform on the site. He required the help of trained observers to identify the many recognizable elements present in the soil besides artifacts, for example, pollen and botanical or geologic components. The more the field worker understood about other diverse but tangential areas, the greater the amount of information that could be extracted from the material. Specialists, like physical anthropologists, or even those who could perform the new dendrochronology techniques, became further integrated into the team.

The more permanent projects established their own laboratories for analysis and preservation. Those that did not forwarded their material to other major repositories like the Smithsonian. The Institution's new acquisitions then made their way through the extensive laboratory system located there. The very complexity of the broader understanding not only enlarged the fieldworkers' core of knowledge but also increased the need for better laboratory facilities.

Another element peculiar to the new archaeologic training and procedural approach did not lend itself to quantification. No longer were archaeologists isolated from surrounding political conditions, as they had been to a much larger degree in the nineteenth century. The ability to carry out academic archaeology resided in the mind and the judgment of the chief archaeologist in the field, as represented by the examples set by Strong in California, Stirling in Florida, and Kelly in Georgia. All three men approached their challenge in a similar fashion; but how different the natural and political environments in which they found themselves!

The New Deal archaeologist faced two problems when he entered an area. In what general area would digging begin, and then, exactly where? The likes of Stirling had to study all the extant information about an area, both archaeological and historical, and then establish a focus. One could find Indian remains just about anywhere in the river systems in North America, quite literally, but how to find the position with the most revealing data? Once a site had been determined, the director surveyed the area, followed by site testing and the selection of one or several spots. What methods would be best to explore a mound: trenching or layering? Such judgment, partially intuitive in nature, could not be culled from books or field manuals. The biases of the individual site leaders proved decisive here, as some archaeologists often showed interest in one culture almost to the total exclusion of others.

Unique circumstances also saddled site archaeologists with social and political responsibilities; not infrequently, they played the roles of both diplomat and negotiator. No hard and fast rules of protocol existed to resolve problems in this realm. By way of contrast, Strong confronted relatively few of the kinds of difficulties faced by Stirling. The latter had to deal with local politicians, labor shortages, and hostile local museum personnel. Each unique situation called for abilities not readily communicated in a methods class. Daily, hourly, or seasonal conditions dictated many of the decisions. Changes in digging methods resulted when fewer available numbers of labor personnel appeared than originally anticipated. Rapidly rising waters in a dammed valley also altered circumstances. Unusual situations, unofficially reported, arose when crews had to harvest a farmer's crops before he would permit a dig to commence. Archaeology, New Deal style, mandated that its professional practitioners be individuals of many talents.

One crucial fact was established by the end of 1934: properly supervised and conceived archaeological projects successfully employed large pools of unskilled workers and provided real, although more limited, opportunities for a lesser number of skilled individuals. Single archaeological expeditions became a vehicle for the employment of over 1,500 men. Although site selection at first favored more mild and temperate weather zones, short-term relief projects followed that utilized human and material resources from other parts of the country. The South, though, became a primary target for this type of relief. The indigenous, heavily impacted Southern economy could little afford to provide much, if any, relief for the rural poor. Workers there, largely unskilled or semi-skilled, subsisted as close to the poverty and starvation level as any group in this nation's history. The nature of these operations required a great deal of hand labor. This made them ideal for the unemployment problems endemic in other rural areas.[102]

An incredible number of archaeologically important sites existed that merited development and research. Indians had inhabited the great river systems of middle and southern America for thousands of years. In some sectors, the heavy demand for hand labor, a persistent need for relief, and the fact that many sites stood remote from urban areas meant that archaeology readily sustained itself throughout the 1930s.

The expansion of archaeology paralleled the explosion of government programs created to respond to the debilitating effects of the Depression. State agencies and universities sustained several of these efforts. For archaeological relief, the Civil Works Administration established a precedent for subsequent operations with respect to specified time constraints. Short-lived surveys and expeditions, usually well defined in intention and goals, became the norm. This idea proved fruitful later. Agencies subsequently employed greater overall numbers of individuals than if crews had been hired for extended periods. Yet application of this philosophy also evoked criticism.[103]

While relief expeditions effectively assisted the general population, they also reaped benefits for members of the academic community. This came about primarily because the Smithsonian established itself as the clearinghouse for all relief archaeological projects during the CWA. This continued its traditional role within government circles and to some extent within the archaeological profession as well. The Smithsonian initiated projects in the Civil Works Administration. It also specified and administered certain employment policies and practices. Later, as these powers declined with delegation of duties to other federal entities, the Institution still performed consultation services for federal agencies. A degree of disciplinary standardization continually evolved as a result of the reliance on professionally trained staff and the declining numbers of noteworthy amateurs used in archaeological projects. The Smithsonian sought recommendations from other academic professionals if members of their own staff could not perform needed tasks. Those chosen inevitably were former students or colleagues. In effect, the universities and colleges supplied the leadership for the relief operations, but the influence and the prestige enjoyed by the Smithsonian was critical in establishing this precedent. Commensurately, these kinds of projects found favor at the higher administrative levels of later relief agencies, especially when the Works Progress Administration classified archaeology under the Women's and Professional Division. Women directors like Florence Kerr and Gay Shepperson sustained the cause of archaeology.

The Civil Works Administration, as did its successor agencies, publicized these projects through the media. The archaeological community consequently received a constant, widespread, and ever-growing exposure to the public at large. The Works Administration carefully utilized news releases to call attention to the digs and what they accomplished, both economically and scientifically. With appreciable cumulative effects, beginning slowly with the CWA and increased by the WPA, information about American archaeology appeared in the papers on daily basis. Contemporary archaeology now vied for the national attention previously directed to the colonial American heritage. Both relief administrations simultaneously attempted to counter derisive commentary, which alleged that relief conditions precluded competent archaeology, especially when untrained personnel labored.[104] Most criticism generally emanated from those people who ran small, "unscientific" excavations and jealously protected their "islands of exploration."

The interest aroused by relief archaeology went beyond good public relations. Tourist activities benefited as well, particularly with the Macon digs. The Macon Merchants Association wrote in February 1934 to Secretary Wetmore, "Local Indian Mound projects have greatly stimulated trade and have given work to 300 men well above the grade of common labor."[105] The dig directors aided this process by speaking at a variety of functions and welcoming visitors. Directors like Arthur Kelly in Macon even conducted night classes for the workers.[106] Relief archaeology, in another sense, functioned as an alternative to museums; people visited the laboratory with the digs in operation. Archaeology became a discipline (defined as a science by many of its professional practitioners) both accessible and comprehensible to the lay members, more so than astronomy, itself another source of popular interest. Neither initial introduction nor further investigation required more than a visit to an excavation or initiation into the "rites" of digging. More detailed astronomical observation, on the other hand, necessitated the purchase of ever more expensive telescopes. Exposure to the phenomena did not entail observation at a great distance either. Kelly informed Wetmore in December 1933, "Certainly the degree of public enthusiasm and interest is unusual. Ford and I have never seen anything like it anywhere."[107] The people who worked the excavations themselves constituted another source of popularization. Site reports document the support and interest that existed, despite the individual's absence of formal training in this subject. Positive, warm remembrances lasted well beyond the 1930s. James Ford communicated the noteworthy degree of optimism generated by such experiences to Frank Setzler in December 1933:

> You really must come down while we are working and see how a CWA project might be run. There is nothing to compare with it in the history of SI archaeology. The only thing we can complain about is the weather.[108]

In short, the Civil Works Administration marked the beginning of a new age in American archaeology.

In the space of fourteen years, the discipline matured from a very fragmented subdiscipline to one that steadily moved toward coalescence. The activity of the State Surveys helped to facilitate at least some degree of communication between professionals by means of publications or regional conferences. Later organizations would reflect further maturation national in scope. The Smithsonian's ethnological cooperation program helped to funnel money to interested agencies. However, these changes became most pronounced when one man, William Webb, through both individual and cooperative effort, and backed by the prestige of the Smithsonian Institution and the National Research Council, campaigned to continue projects under the auspices of the CWA and the Federal Emergency Relief Administration. Academically trained personnel gained critical field experience, earned salaries, and faced a greater prospect for later securing more permanent academic employment.

The Civil Works Administration projects set a precedent for the federal relief programs that followed, whose novel yet ambiguous characteristics fostered further organizational and administrative opportunities for university-based archaeologists outside the Smithsonian. The avowed notion that professional archaeology was a university-oriented function became a legacy of this agency as well. The Civil Works Administration and its successor programs set a standard for training the next generation of leaders in the archaeological community, individuals whose attitudes and approaches were in good part forged by their experiences in the relief projects.[109]

The Smithsonian filled another important role by asserting the importance of publication. Through that body, large amounts of information more readily found their way into print; this new data created a momentum of its own. Pressure developed to reassess prior conclusions and ideas. In turn, new approaches migrated beyond arbitrary political boundaries.

As the Civil Works Administration ended, one fact stood out: key Smithsonian personnel remained convinced of the future viability of such projects. Setzler reflected on this experience:

> This experiment convinced us that with a sufficient number of trained assistants and supervisors, it would be feasible for archaeologists to use large crews of relief labor. From the standpoint of the relief agencies, archaeological explorations were made to order, for they provided immediate work for numerous unemployed people; they used more than 90 percent of the allotted funds for labor, since the necessary tools and other materials cost very little; and they produced results of a scientific and educational nature.[110]

With the staff of the "nation's attic" thoroughly convinced about the efficacy of this approach, they but awaited another opportunity to implement it further. The persistence of the economic doldrums assured that this would occur.

Chapter 3

Federal Emergency Relief Administration: A Year of Cooperative Transition

By the time the Civil Works Administration ceased to function, the most severe phase of the Depression seemed to have ended. However, little prospect of immediate improvement appeared on the horizon. Therefore, Roosevelt's administration began to consider more comprehensive and permanent solutions to the problems that plagued the country. One aspect of this more aggressive posture centered on the creation of a more efficient relief program that encouraged "honest" labor, not a continuation of the "dole."[1]

While the Second New Deal underwent reformulation in the minds of policymakers, a span of one year divided the Civil Works programs and the permanent Works Progress Administration: a time of transition. The Federal Emergency Relief Administration remained in operation and continued to oversee a scaled-down public works program. While the Depression-era government marked time, professional archaeologists moved to consolidate their newfound status through formal unification. They also expanded their pool of conceptual knowledge through the continued efforts of the Tennessee Valley Authority. These two themes stand as the most significant ones for the pursuit of this study: the evolution of the field's political consolidation and its increasingly professional research. Still, other forms of work did occur in archaeology but in a milieu more conducive to maintaining regional and local emphasis. Lesser projects returned to state and local control, with limited government funding. The bulk of the federal funding continued with direct relief enterprises, but decentralization in the decision-making apparatus appeared. Reference appendices discuss such matters.

The one-year period from 1934 to 1935 aided in creating the future political circumstances in which WPA archaeology thrived. Strong currents of cooperation were evident. With scaled-back operations, the cost of funding projects was increasingly shared. Additionally, the FERA offered less administrative oversight. In this altered setting, greater flexibility regarding the inauguration and administration of work relief prevailed. The power of the Smithsonian Institution thus declined substantively. Universities, museums, and state agencies, rather than the SI, produced applications for funds and labor and essentially assumed self-direction.[2]

This new climate meant that a non-federal, bureaucratically generated relief mechanism affected decisions regarding implementation of archaeological investigations and the personnel on which they relied. These circumstances ushered in a period when almost any political body in the country could sponsor a project. Nevertheless, promulgation of that procedural modification brought larger portions of the academic community into the archaeological relief business than had been the case under the CWA. It also stimulated the development of cooperative endeavors between institutions studying similar phenomena.

Few Relief Administration projects functioned as cooperative endeavors. These projects did not encourage the migration beyond any region of shared managerial scientific procedures, much less a common classification system. What became apparent during the life of this relief organization, and under the later Works Progress Administration as well, was that institutional (including government, museum, and college) academic archaeology evolved in an interesting manner. Different areas of the country advanced towards disciplinary unification (e.g., emergence of temporal classification systems as well as organizations) at varying rates. Therefore, mere infusion of funds failed to assure uniform dissemination of academically derived methodologies. Other factors played decisive roles.

Viewed in their totality the New Deal programs of the 1930s did not create the foundations of American archaeology. Rather, a fragmented base existed in a plethora of stages of maturity. The involvement of agencies of the federal government in relief archaeology assumed one of two forms. In the first, agencies proposed projects and funded them, including the costs of labor. In the second, agencies paid only for a supply of labor. In both cases, federal activity produced a similar result: it hastened the rate at which archaeology became more politically and organizationally unified. The New Deal, in short, created an atmosphere conducive to the advance of an academic community based largely in the nation's colleges and universities.

In the midst of this fascinating, if often inconsistently implemented, relief world, TVA programs spurred new theoretical and methodological initiatives. Power was divested to the states, thereby reinforcing state and local conceptions of archaeology. Federal money found its way into archaeology through means other than relief agencies. For large power projects like those in the Northwest, the Civilian Conservation Corps, the National Park Service, or the National Youth Authority provided funds.

FERA archaeological cooperative work that engendered further political and professional maturation divides this portion of the study into two realms: that of the Tennessee Valley Authority and the movement toward academic national coalescence. Of particular note, the Smithsonian still participated in this sphere but in different ways. The Institution, having regained funding for expeditionary work through its yearly congressional appropriation (recall Hoover's earlier budget slashes), no longer depended solely on relief monies for digs. However, as a player in the post–field analysis mechanism, the Smithsonian analyzed material forwarded to it by a variety of universities. In addition, Smithsonian personnel also served on the advisory committees instrumental to TVA activity.[3]

By the mid-1930s, federal money had played a major role in the expansion of archaeological investigation. Yet, the research process lacked full management by academic archaeologists. As expected, the community viewed certain research results rather critically. Professional archaeologists therefore intensified their efforts to consolidate political control. This was manifested in regional cooperation and in the impetus toward the creation of a national organization, one that would represent the diverse needs of American archaeology. Dissatisfaction stemmed from low levels of funding, the lack of recognition archaeology received from academic organizations, and the hegemony exercised nationally by cultural anthropologists and other social scientists. Recall also that archaeology was categorized under anthropology and sociology such as with the NRC. All such currents fed the formation of the Society for American Archaeology.

The acknowledged lesser status of archaeology relative to anthropology and other subdisciplinary components like ethnology became unavoidably apparent by 1934. The editors of *American Anthropologist* indicated that due to budgetary constraints the section of the journal that carried news of the archaeological surveys would cease to exist. The National Research Council funded publication of the last year of state surveys, 1934.[4]

Carl Guthe became the driving force in the campaign to establish archaeology in its own right, a discipline that could move from under the shadow of the National Research Council's Division of Anthropology and Psychology, the Smithsonian, and the American Anthropological Association. After all, more academic archaeologists than ever before staffed government, museum, and college institutions, with funding from many sources. Guthe represented the university-based element of the emergent community, one which labored to overcome the fragmented nature of its composition.

The State Survey Committee's stated goal was to "stimulate a greater scientific interest in local archaeology in a few Middle Western States." Soon it became evident, though, that the Committee would also have to act in an advisory capacity regarding other matters, for example, specification of proper techniques and methods for use in "scientific archaeology" (thus establishing a precedent for a later group, the Committee for Basic Needs).[5] The State Surveys Committee then expanded its role into that of a clearinghouse for information on North American archaeology. Its geographical influence expanded as well. Consequently, the need for greater cooperation, between professionals and non-professionals alike, as well as within the academic community at large, became evident, as members became more knowledgeable about each other's activities. Such circumstances prompted Guthe on 28 December 1934, at the annual meeting of the American Association for the Advancement of Science, Section H, Division of Anthropology, to declare that a more effective mechanism to address all these needs merited creation.

The initial response to Guthe's statement involved the organization of a subcommittee to evoke increased cooperation between local societies; short-lived, it produced no tangible results. Clearly, professional archaeology required a permanent committee, one not dependent on foundational support. Guthe and the other professionals also deemed vital more useful means to accommodate, yet effectively govern amateur archaeologists. These academics did not want to alienate or otherwise isolate the non-professionals; that circumstance only encouraged pot-hunting. Instead they advocated education, and even control, of such amateurs by including them in the new framework.

While this attempt to work through the organizational structure of the American Association for the Advancement of Science proved fruitless, a small coterie of individuals, Carl Guthe again included, pursued other alternatives. An *ad hoc* meeting of this group at this same affair in Columbus, Ohio, witnessed the preparation of a plan. They then mailed out a prospectus to 192 professionals by spring 1934. Eighty-five responses were received. The survey elicited an overwhelming reply in favor of the creation of a novel organizational structure. A committee, elected to write a constitution and by-laws, consisted of A. V. Kidder, A. L. Kroeber, and Frank H. H. Roberts, Jr.[6]

Archaeology had embarked on the path that eventually led to a measure of independent standing for the discipline. Still, Guthe carefully emphasized that archaeologists had not assumed an identity totally separate from anthropology. Instead, they desired to erect structures for the enhancement of cooperation amongst their brethren. They also hoped to bolster archeology's growing public image, and in turn to further educate the public. Vandalism, looting, and pot-hunting might be controlled if the public understood more fully how scientific archaeology should be conducted. And, of course, archaeological work would become the privy of the professional; the public would defer to those so trained.[7]

The time is of some interest. It was late in the day after a full day of attending a conference. Hence, it took only an hour to debate the creation of the new society. After a short discussion, the constitution and by-laws were approved. Guthe stated succinctly, "It had taken just one hour to bring life to the Society for American Archaeology."[8] Its designated journal, *American Antiquity*, commenced publication in July 1935.[9] The new society, a "reincarnate" of Guthe's State Archaeological Surveys Committee (it persisted until 1937), gave new voice to the professional community. An opportunity to resolve the well-entrenched amateur–professional identity problem now arose too.

The professional personnel who comprised the officers and council of this body, names familiar to this study, stood out as acknowledged leaders from museums and colleges. The structure of the new organization relegated power to those who already exercised some control of the avenues of status and other important considerations in the discipline.[10]

In the same year that witnessed the birth of the Society for American Archaeology, the most significant archaeological work of the New Deal era took place in the Tennessee River Valley. Two huge expeditions brought archaeologists together in a fashion never before observed in the history of the discipline. Analyses of the material collected indicated both the degree of specialty growth in archaeology and how important consensus on methodology and procedure had become. William S. Webb oversaw the exploration by Federal Emergency Relief labor of the areas behind Norris and Wheeler Dams. Cooperative in scope, this project fit conceptually into the dynamics of this transition period. The digs epitomized the best and the largest example of the growing cooperative environment that came to the fore in this year, both regionally and nationally.

Information from these explorations subsequently was published in two lengthy Bureau of American Ethnology *Bulletins* in 1938 and 1939. Archaeology's spokesmen consistently expressed concern that information gathered from excavations should find its way into print in a reasonable amount of time. The activist posture of the Smithsonian in the Civil Works Administration insured that at least general narratives of the data emerged. The university-trained staff of the SI, disciplined to research and publish, matched this behavior with the long-standing tradition of the Institution to act as a repository of scientific information. In the case of the Tennessee Valley Authority, officials receptive to the call subsidized the cost of printing. However, in other instances, and after archaeological projects branched out from the control of the Smithsonian, publication did not always result, especially during the WPA period. Archaeological leaders correctly perceived that information must be analyzed and disseminated to the larger community before new and larger interpretations and classification systems might materialize. Recall the historic development of American archaeology, where professionals, isolated in institutions around the country, experienced little communication and sharing of data before 1900. The identifiable result was that no large taxonomic or classificatory systems emerged.

Norris Dam stands on the Clinch River, some eighty miles above the confluence with the Tennessee River and seven miles below the mouth of the Powell River. Upon completion, the new Norris Lake covered about fifty-three square miles and considerable Indian habitation. The initial archaeological survey established twenty-three sites made up of twenty earth mounds, nine stone mounds, four village sites, and seven caves. Twelve of the sites housed burials and seventeen contained associative structures.

The University of Tennessee received artifacts. Skeletal material was then forwarded to the University of Kentucky. The Ceramic Repository at the University of Michigan examined the samples of potsherds. The Universities of Chicago and New Mexico analyzed wood samples in a dendrochronological study.[11]

The nature of this analysis illustrates the growing importance of specialization to archaeology. Webb certainly did not possess the skills to deal with all the aspects of physical anthropology, or even all of the ceramic artifacts; thus, he depended on various experts, a trend in modern science. The fact that such individuals existed testified to the creation and dissemination of such knowledge within universities. This phenomenon drew from the large institutional precedent set by Guthe and Fay-Cooper Cole when they organized the Ceramic Repository in Ann Arbor. Both during and after completion of the digs, the continued reliance on specialists to supplement research that originated in the Tennessee River Valley persisted. The amount of information gathered would overwhelm one or two archaeologists. They could not monitor it and simultaneously direct the field operations. Physical anthropologists and ceramists usually accompanied the expedition, while paleontologists analyzed the flora and fauna.

It should be pointed out that specialization had been present in anthropology since its inception: archaeologists, ethnologists, linguists, and physical anthropologists dated back to the nineteenth century. As further cultural definition and sequencing delineation emerged, more-defined specialization came to archaeology, both internally and from disciplinary experts. Accordingly, the flow of information increased and the dimensions of supportive scientific investigations from disciplines such as biology expanded. Archaeology became more exacting.[12] The whole process was accelerated by the scope of the relief projects, especially those in the Tennessee Valley. They resulted in published analyses of a substantially different nature. Additionally, separate individuals authored distinct portions of the publications. By way of contrast, only one or two authors had penned the many works from the nineteenth century. Naturally, the depth of that investigation was not nearly as comprehensive. The early anthropological work addressed only contemporary features of extant Indian culture, for example, linguistics. The archaeologists of the 1930s, far more sophisticated in their research, dealt with a much broader, complex domain.

Webb's conclusions regarding Norris Basin reflected a rather limited scope. He attempted to compartmentalize cultural traits into two complexes designated "large log town house complex" and "small log town house complex."[13] Webb, comparing his information to that in previous studies, found some correlation, but not enough to suggest a recent Cherokee habitation, although later interpretive syntheses did just that.[14] W. D. Funkhouser analyzed and reported on the skeletal remains and deduced, based upon evidence gathered from surrounding areas, that the inhabitants of the Norris Basin represented the "same stock and probably closely related groups farther north and west in the Mississippi Valley."[15] James B. Griffin examined the ceramic remains at the Michigan repository, the composite information forming the basis of his doctoral dissertation at the University of Michigan in 1936. In almost 100 pages of the Norris Basin *Bulletin*, Griffin analyzed the pottery retrieved from ten of the sites. He concluded that it was difficult to compare the results from this find with others. He lacked a complete comparative framework to create a definitive classification system. However, he did find, after a detailed inspection, that these shards, while similar to those of the Upper Tennessee River, known as Cherokee, showed several important differences as well: decoration of the jar shapes and new pottery types, such as with the water bottles, exhibited significant variation. Hence, he remained tentative in his conclusions and waited for further postwar studies.[16]

The first of these River Valley expeditions revealed the great potential of large-scale work. The new information created new and more demanding classification measures, as evidenced by Griffin's need for more data. The differences in approach and scope between Webb and Griffin were revealing. Webb depended on previous information on Kentucky and Tennessee to reach conclusions, whereas Griffin envisioned a potentially larger cultural area. The use of ceramics remained the key classification agent to determine a larger, if not clearer, representation.

The cooperative nature of the River Valley expeditions paralleled concurrent changes in the larger archaeology community. The Ceramics Repository operated as a central research facility with a combination of academic institutions working together to achieve the common goals of investigation, analysis, and publication. Not unexpectedly, the various academic specialty areas of archaeology interfaced more efficiently with each other than archaeology did with its parent discipline of anthropology.

The interpretive work of James Griffin in the Tennessee areas reflected other changes in the field. Eastern archaeology would witness the creation of a regional interpretation for the first time. A. V. Kidder and others had already erected a classification structure in the Southwest. Now Griffin began to lay the foundation for broad interpretation in the East. Here, the young professional archaeologist James Griffin, not the field leader William Webb, predominated. Webb's leadership served one vital function: he created and managed a unique environment conducive to the acquisition of material and information. The special skills of Griffin then translated that data into a far more ambitious conceptual framework than any that had preceded it. Griffin, a trained ceramist with a keen analytical mind, was extremely adept at cataloging. The other sites in the South which evolved during the Civil Works and Emergency Relief Administration brought forth much less comprehensive analysis.

The cooperative nature of this cross-institutional investigation also led to the introduction of dendrochronology into the southeastern study area. Webb believed this could be one of the more important results of the study in Norris Basin. The impetus to apply this technique had started earlier at Chicago. There, Fay-Cooper Cole and ring-dating expert Florence Hawley of Arizona in early 1934 had proposed a study of tree-ring dating in the East. The State Survey Committee under Guthe published this information and described the Chicago request for wood samples. While the dendrochronologic dating attempt at Norris Basin proved inconclusive due to insufficient information, the attempt to spur dissemination of this technique had begun.[17]

The second major project directed by Major Webb concerned Wheeler Dam, planned and named by the TVA for General Joe Wheeler, a Confederate and American general. Located at the upper end of Wilson Lake and formed by Wilson Dam, the Wheeler Dam area extended to Buck Island above Guntersville, flooding a section of the Tennessee River in northern Alabama some 80 miles square. A primary archaeologic survey by Walter Jones of the Alabama Museum of Natural History between 1932 and 1934 had located some 273 sites, the most important being 26 in Lawrence County, 36 in Limestone County, 11 in Lauderdale County, 50 in Morgan County, 37 in Madison County, and 77 in Marshall County, with targeted sites in each area.[18]

This large endeavor required another cooperative effort. Guthe's State Survey Committee publicized the request for supervisors.[19] The Alabama Museum became involved and loaned equipment and personnel, notably David DeJarnette. Artifacts came to the Alabama Museum, while shards again went to the Ceramic Repository in Ann Arbor. Once more, various universities provided the field party supervisors, their selection based not only on previous academic training in the latest archaeological methodology but on field experience as well. They included: Robert Adams, Elliot Davis, Kenneth B. Disher, James R. Foster, Bennett T. Gale, D. W. Lockard, Horace Miner, Robert D. Morrison, J. J. Renger, Alden B. Stephens, and James W. White.[20]

The analysis of the Wheeler expedition offered more substantive information than that gained from Norris Dam. This time, Webb projected more expansive conclusions by proposing a classification for a new cultural complex: a Copper-Galena (Copena) complex in northern Alabama, based upon the Midwestern Taxonomic System.[21] Later information and interpretations by James Ford, Gordon Willey, and James Griffin altered this conclusion.[22]

W. D. Funkhouser analyzed and reported on the physical remains, as he had done in the Norris study. The skeletal material once more was sent to the University of Kentucky.[23] James Griffin, after an examination of the shard evidence, determined the type of temper as the single definable characteristic, but the surface finish, decoration, and shape utilized distinguished culturally complete ceramic types: shell-tempered, fiber-tempered, sand-and-grit-tempered, and limestone-and-clay-tempered. Due to the distinctiveness of each type, no evidence of cultural intermixture appeared.[24]

A classification system implicit to the data had materialized, which could then be compared to categorized evidence from other areas. The extensive materials taken from areas slated for destruction made this possible, although in a subtly ironic way. Archaeologists had slowly, albeit at a more rapid rate than in the nineteenth century, taken increasing amounts of evidence from small, focused expeditions. Griffin, able to apply his specialized training, provided a taxonomic, analytical order to it. A more precise chronological specification now remained in need of construction.

Several substantive achievements resulted from just one year's work. A level of cooperation never before experienced in the history of the subdiscipline had taken place on two large fronts. The reliance on the use of specialized skills multiplied in direct proportion to the scale of information retrieved. The cooperative thrust involved schools throughout the country and within the region. The hiring methods adopted by the Authority and Webb, following the lead already set by the Smithsonian, bolstered academic archaeology's position in the relief structure. As the program of salvage archaeology fused into government circles, a procedure for rapid exploration developed: a preliminary survey, followed by an analysis of the information from key sites, then an expanded research network incorporating archaeologists and graduate students into the process, coupled with the utilization of their laboratory facilities.

The initiation of salvage archaeology projects also raised, for some, troubling questions. Given the severe time strictures facing field supervisors, combined with the unique circumstances that surrounded the work, could "good" archaeology result? Obviously, that term was a rather relative, ambiguous one. Specialists analyzed archaeological material with the finest laboratory systems then available. However, what of the procedures in the field? The very essence of salvage archaeology created narrow temporal and geographic limitations. The available time left to complete the digs became an inverse function of the speed at which the water level rose; appreciably less area in which to work remained. Still other unusual pressures faced the director. Was the survey adequate and sufficiently extensive? Which areas should be selected after the preliminary survey? Which criteria for the selection were given priority? The sites with the most remains, the most ceramics, an identifiable culture, or simply those that were most familiar to the director? And what should be taken from the dig? By raising these queries, archaeologists indicated a concern about the viability of relief work. Of course they could do nothing and lose all the evidence. In short, archaeologists still grappled with the problem of how best to define the precise parameters of scientific archaeology in a time and place not of their own choosing.[25]

Whatever the nature and source of criticisms or doubts about the early relief projects, first under the Civil Works Agency and later under the Federal Emergency Relief Administration, by the spring of 1935 two facts stood out. First, this experience convinced both the emergent professional community and important elements within the federal government about the viability of effective, useful relief archaeology. As such, this endeavor would endure in the Tennessee River Valley. Secondly, this form of archaeology would spread to other parts of the country as needed. The next stage of the maturation of a discipline, replete with its growing political sophistication, evolved as the result of the almost permanent relationship formed between American archaeology and the federal government, a phenomenon inaugurated under the Works Progress Administration from 1935 to 1942.

A CREW AT WORK FILLING IN EXCAVATED AREAS AT PICKWICK BASIN, ALABAMA, 1940. (Photograph by Wayne W. Kraxberger courtesy National Anthropological Archives, Smithsonian Institution.)

A TYPICAL MIDDEN PIT AND CONTENTS, LAWRENCE COUNTY, ALABAMA, 1940 This photograph captures the essense of New Deal archaeology: helping the poor blacks and whites and recovering remains. (Photograph by Steve Wimberly courtesy National Anthropological Archives, Smithsonian Institution.)

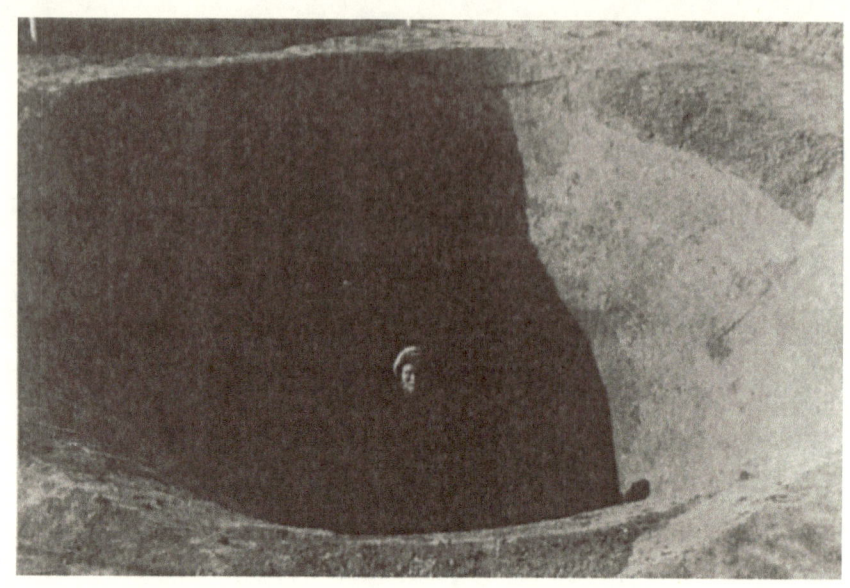

AN ABORIGINAL CELLAR, OR A MIDDEN PIT NINE FEET DEEP, LAWRENCE COUNTY, ALABAMA, 1940. (Photograph by Steve Wimberly courtesy National Anthropological Archives, Smithsonian Institution.)

THREE-FOURTHS OF SITE EXCAVATED AND BACK FILLED, WHEELER BASIN, LAWRENCE COUNTY, ALABAMA, 1940. (Photograph by Steve Wimberly courtesy National Anthropological Archives, Smithsonian Institution.)

Large Steatite Bowl Inverted over Portion of Burial No. 21, Wheeler Basin, Lawrence County, Alabama, 1940. As the older gentleman crouching over the remains shows, New Deal archaeology placed no age restrictions on field workers. (Photograph by Steve Wimberly courtesy National Anthropological Archives, Smithsonian Institution.)

General View of a Wheeler Basin Site under Excavation, Lawrence County, Alabama, 1940. (Photograph by H. Summerfield Day courtesy National Anthropological Archive, Smithsonian Institution.)

TWO MEMBERS OF LOCAL SOCIETY FOR GEORGIA ARCHAEOLOGY, MACON, GEORGIA. (Photograph courtesy Ocmulgee National Monument, Macon, Georgia.)

A FIELD SHOT OF MOUND D EAST OF MACON, GEORGIA
The photograph shows the town of Macon in the background. (Photograph courtesy Ocmulgee National Monument, Macon, Georgia.)

MOUND CAT DIG OUTSIDE MACON, GEORGIA
Lots of river sediment and overlay made these
archaeological projects labor intensive.
(Photograph courtesy Ocmulgee National
Monument, Macon, Georgia.)

ARCHAEOLOGICAL STAFF AT SNAKETOWN, MARCH 1935
Left to right, Emil W. Haury, Julian Hayden, Evelyn Dennis, Irwin Hayden,
E. B. Sayles, Nancy Pinkley, Eric K. Reed, J. C. Fisher Motz.
(Photograph courtesy Arizona State Museum, University of Arizona, Tucson.)

WILLIAM SNYDER WEBB. (Photograph courtesy Division of Special Collections and Archives, University of Kentucy, Lexington.)

THE LOUISIANA STATE UNIVERSITY-WORKS PROJECTS ADMINISTRATION

ARCHAEOLOGICAL SURVEY PROJECT

Introduction

Archaeology is the study of past civilizations, both simple and elaborate. Such civilizations as those possessed by European cave men, ancient Egyptians, Greeks, Romans, Aztecs, mediaeval Frenchmen or Louisiana Indians are all subjects for investigation. The archaeologist studies past civilizations so that he may learn and record their histories. In cases where complete and accurate written records describing a civilization have been preserved, archaeological studies are not necessary, but in many instances such written histories have not been kept, and it is these historyless civilizations which are of particular interest to archaeologists.

We shall elaborate upon this theme, using as our prelude, the writings of Dr. Carl E. Guthe of the University of Michigan. "The archaeologist," says Guthe, "is the historian who attempts to understand the simpler civilizations, using as his source material the unintentional records the people left behind them. Obviously it is difficult to translate such fragmentary records as are found in refuse heaps, ruined buildings, broken vessels, forgotten graves, and the discarded ornaments, implements, and utensils, yet the archaeologist must be trained to read this story as it occurs in the ground.

"The technique is that of the documentary historian. The site itself is the document; the layers of deposits unintentionally laid down may be considered the lines of the text; while the object themselves constitute the words. In order to read the story, the archaeologist must carefully preserve the relative positions of the materials in their proper line and with relation to one another upon that line. Just as in a document, if these words are disarranged and lifted from their context the story is lost. The great worry of all archaeologists is that they may fail to see the true significance of the record as they uncover it. As the trowel and the shovel move the earth the record is destroyed

INTRODUCTION TO REPORT OF THE LOUISIANA STATE UNIVERSITY-WORKS PROGRESS ADMINISTRATION ARCHAEOLOGICAL SURVEY PROJECT. The Louisiana survey, the first WPA archaeological enterprise, set the precedents for all subsequent WPA-sponsored digs around the United States. (Reproduction courtesy National Anthropological Archives, Smithsonian Institution.)

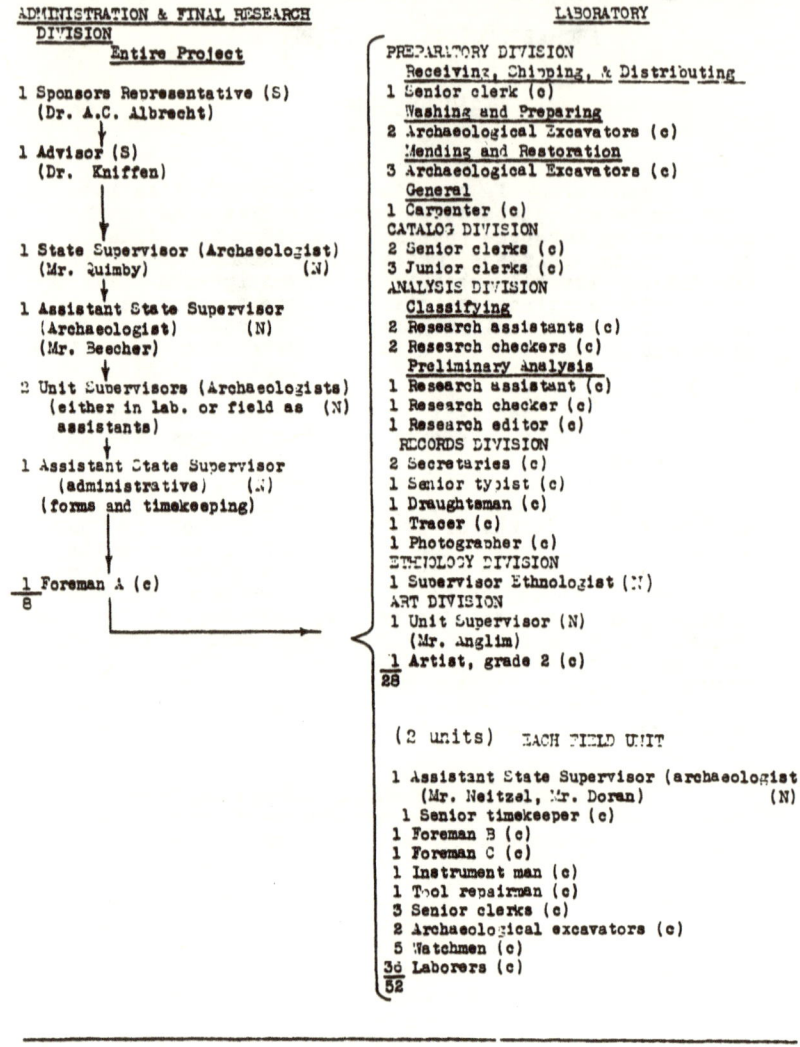

DIAGRAM OF PROJECT ORGANIZATION FOR THE LOUISIANA WPA ARCHAEOLOGICAL SURVEY. WPA archaeological surveys around the United States adopted the Louisiana organizational scheme. This chart shows the division of labor and complexity of the WPA archaeology. (Reproduction courtesy National Anthropological Archives, Smithsonian Institution.)

ARCHAEOLOGICAL SURVEY OF LOUISIANA
WPA-LSU

PROGRESS RECORD FORM   Date _____

Excavation _____   Site _____

[grid for recording data]

Discoveries: Labor:
                                                          Present Absent

Fea. Nos.			Burial Nos.			Fd. Nos.		
Start	Cont	Comp	Start	Cont	Comp	Start	Cont	Comp

No. added _____ _____
Unskill. _____ _____
Intermed. _____ _____
Skilled _____ _____
Prof. _____ _____
Sup. _____ _____
Totals _____ _____
Total assigned _____

Photographs: From Pack ____, Neg. ____ to Pack ____, Neg. ____ inclusive.

Specimens or equipment shipped, received: _____ by _____

Weather: _____ No. hours worked today _____

Remarks: See reverse of this sheet.

                                  Reported by _____
                                  Checked by _____

PROGRESS RECORD FORM FROM THE LOUISIANA WPA ARCHAEOLOGICAL SURVEY Project directors around the country routinely filed this form to report the latest developments in their digs. (Reproduction courtesy National Anthropological Archives, Smithsonian Institution.)

LABORATORY PROGRESS RECORD FORM

Date:_____

No. of hours worked:_____

Total workers assigned:_____

Cartons of field material received:_____ Materials washed_____ bags.

No of washers:_____ Find Control entries:_____ Boxing & labeling_____ boxes.

No. of clerks:_____ Catalog entries made:_____ Clerks, entering:_____

Pieces of material numbered with
catalog & site numbers, & counted: TOTAL COUNT:_____ Clerks, numbering:_____

  Sherds:_____ Teeth:_____ Ground Stone:_____ Misc. Clay:_____

  Bones:_____ Chipped Stone:_____ Other Stone:_____ Other Misc._____

Pieces restored:_____ boxes. Restorers:_____ .

Artifacts restored by plasterings:_____ Other restorations:_____ Restorers:_____

Negatives developed:_____ Prints made:_____ Prints & negatives cataloged:_____
Photographs pasted on cards:_____
Plate drawings made:_____ Field drawings copied:_____ F.drawings enlarged:_____

Carpenter's summary of day's work:_____

Other misc. work such as packing & shipping of materials:_____
_____

Field forms copied:_____
Pages of field notes typed:_____
Pages of manuscript typed:_____
Letters typed:_____
Other typing:_____

ANALYSIS:
Estimated no. of pieces handled in primary sorting:_____ Sorters:_____
  Remarks:_____

Estimated no. of pieces handled in primary classification:_____ Classifiers:_____
  Remarks:_____

Estimated no. of pieces handled in final classification:_____ Classifiers:_____
  Remarks:_____

SUPERVISORY:
Dioramas in prospect; progress estimated in fractions for each:_____
  Remarks:_____
Material analysis; progress estimated in fractions for each site:_____
  Remarks:_____
Plates of drawings completed:_____
Plates photographed:_____
Publications in prospect; progress estimated in fractions for each:_____
  Remarks:_____
Labor present: No. added:_____ U.____ I.____ Sk.____ P.____ Su.____ Totals:_____
Labor absent: :_____ U.____ I.____ Sk.____ P.____ Su.____ Totals:_____
Reported by:_____ Approved:_____

Remarks: See the reverse side of this sheet.

LABORATORY PROGRESS RECORD FORM FROM THE LOUISIANA
WPA ARCHAEOLOGICAL SURVEY. The WPA required survey directors to keep and turn in
strict records of the collection and analysis of archaeological artifacts. (Reproduction
courtesy National Anthropological Archives, Smithsonian Institution.)

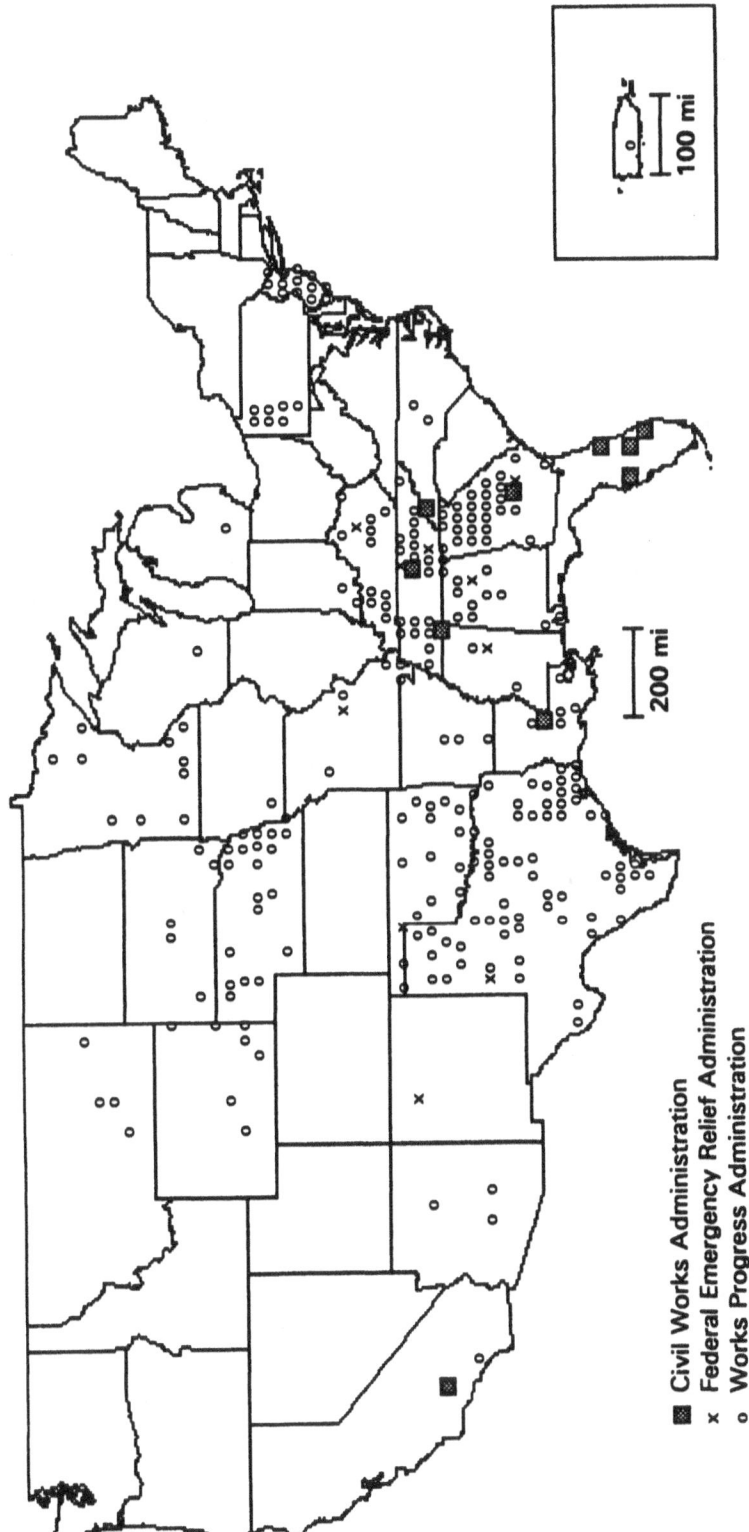

New Deal Relief Archeology Sites

Chapter 4

WORKS PROGRESS ADMINISTRATION: A NEW ORGANIZATIONAL ENVIRONMENT

Franklin D. Roosevelt proposed to Congress in January 1935 a huge program of emergency employment for the jobless.[1] After negotiation of several issues centering upon what should constitute an acceptable wage scale, Congress authorized the Emergency Relief Appropriation Act of 1935. The largest single expenditure in the country's history to that time gave Roosevelt the power to distribute the funds as he deemed necessary. The President opted to channel the largest portion of the nearly $5 billion allotment to Harry Hopkins and the WPA. Given the pressing nature of the nation's economic plight, Roosevelt's actions allowed Hopkins to follow his straightforward approach: put as many people to work as possible with the available money.[2]

The Works Progress Administration expanded the administration's emergency policy toward the Great Depression. The ability to respond rapidly to individual problems, be they agricultural or housing, for example, could forestall further hardship. Although conceived to be of longer duration than previous alphabet agencies, an atmosphere that created confusion, overlapping authority, conflict, ambiguity, and decentralized decision-making prevailed. Yet, the Works Progress era stood out from the period that immediately preceded it. The Civil Works Administration and the Federal Emergency Relief Administration, short-lived by design, had not functioned long enough to merit restructuring to meet or adjust to new demands.

Distributed throughout various levels of government and among several agencies, the vast WPA works agency decision-making apparatus peculiar to archaeology lacked cohesive, fully centralized form. State and local organizations arranged labor distribution, some funding, determined eligibility for relief work, and supplied the labor crews. Universities, museums, or other sponsoring institutions policed themselves and insured adequate scientific standards. Hence, neither a thorough conception nor practice of scientific archaeology materialized. Instead, a more decentralized environment for attempted sufficient monitoring of relief-era archaeology held sway. At the same time, however, more institutions were able to participate, thus spurring further investigation around the country.

Early administrative relief policy became subject to review and revision after the programs existed for some time. While the original emergency relief needs, subject to rapid, often *ad hoc* implementation, continued, procedures became more carefully structured. In short, the bureaucratic apparatus within which archaeology operated evolved through time. Complaints pertaining to matters such as labor allotments, helpful suggestions from local, state, and participatory groups, coupled with advice from other sub-governmental agencies, led to the reformulation of original administrative approaches. Still, early on, scientific quality depended on those personnel in the field. Later modifications produced two important results that affected academic archaeologists. First, altered guidelines acknowledged the primacy of university-based archaeology. Second, now in a stronger position, archaeologists sought further political control, both over the discipline and within government circles, to insure higher standards of academic practices.

This crisis-oriented administrative apparatus, in flux as it was, significantly affected not only the course of New Deal archaeology, but also the ongoing movement toward professionalization, the latter characterized by the marked ability to act in a cohesive political manner. Several important questions then arise, which govern the following discussion. How and why did the Works Progress Administration assess and balance the distinct needs of relief workers and archaeologists? In what ways did factors like severe limitations on available time for excavating, conflicts over control of monies, changed administrative structures, and lack of uniformly applied scientific standards affect early WPA relief archaeology? Did these ideas and policies change through time or influence later relief archaeology efforts? What was the essential impact of any Works Progress Administration attempt to institute higher (academic) scientific archaeological standards? How did this relate to the emerging managerial procedure and the continued quest for control of the discipline by academic archaeologists? Within this unique environment, how did certain ideas and strategies develop at the field level that dramatically affected the evolution of the community, both politically and scientifically? Which political developments resulted in coalescence and greater control by academics? Which ones followed from planned efforts by archaeologists? And did any emerge as the product of economic and administrative evolution and process? Since, in part, many of the actions of university-based archaeologists challenged the Smithsonian's lofty position, when and why did the SI either support or repudiate proposed reconfigurations of policy?

Upon its formation, the Works Progress Administration immediately confronted a delicate situation. On the one hand, it had to address the expectations and needs of the unemployed. At the same time, federal relief aimed not to compete directly with the private sector. Additionally, the costs of the projects could not be inordinately high, nor could they displace extant federal programs. Dictated by those constraints, many programs emerged of a makeshift nature. Despite such limitations, Hopkins as usual proved to be ingenious. Project accomplishments included hospital construction, airport improvement, schools made earthquake safe, ski lodges erected, art classes convened, a Braille historic records inventory completed, and creation of the Federal Theater Project, the Federal Writer's Project, and the Art Project. The WPA thus further institutionalized the Hopkins' relief ideology: everyone deserved an opportunity to earn a wage irrespective of their skills.[3]

Harry Hopkins did not ignore the lessons of the past, either. The Civil Works Administration had been subjected to considerable criticism. Political improprieties, economic fraud, and inefficiency had surfaced, in part due to its rapid implementation and a need to utilize state and local political frameworks. To control more carefully the dissemination of power and the allotment of money, the Works Progress Administration adopted a different strategy. The WPA, viewed as a more aggressive program of indeterminate length, would enable the unemployed to support themselves until the economy achieved some degree of renewal. The agency placed a limit of eighteen months on continuous relief service in order to ensure that as many people as possible would benefit from this effort. It was also hoped that the designed time stricture would prevent the relief programs from extending into a period of recovery. Still other checks on excesses pertained to hiring and work practices. For example, persons could be reassigned to other jobs only after a waiting period of thirty days, with work hours fixed at 8 per day, 40 per week, and 140 per month.[4]

The Civil Works Administration and the Federal Emergency Relief Administration functioned by means of large grants allocated to the states. The decision-making process relative to addressing immediate needs (e.g., number of workers needed versus those made available by local officials) filtered down to agencies at various levels. Previously, the vicissitudes of local politics had at times interfered with implementation of public policy. This prompted Harry Hopkins to preclude repetition of such activity. Therefore, the Works Progress Administration (even under its changed name in 1939 to the Works Projects Administration) operated as a self-contained entity. That is, Hopkins concentrated greater power in it, often to the total exclusion of other organizations. Government relief bureaucratic structure had become much more well defined.[5]

In some respects, the Works Administration utilized established accounting procedures. To ensure that monies were expended in the manner intended, the agency relied on extensive accounting paperwork, audit, and, if deemed necessary, intervention. Federal–state cooperation regarding these fiscal matters persisted. Contrasted with the CWA's loosely organized "shadow bureaucracy," tighter federal review became the rule. Projects were subjected to budget and value evaluation. Works Agency district offices, both within each state and at the regional level, monitored the activities of the various programs.[6] Hopkins also employed investigators at large, like Lorena Hickock, who toured the country and scrutinized all types of relief projects and the personnel responsible for them.[7] Approval of *all* applications for relief activity, associated wages, and other expenses first passed through local and district offices. If determined worthy of support, the Washington headquarters then reserved final judgment.

In its pursuit of continued support, then, archaeology confronted an altered bureaucratic structure with two distinctive operational characteristics. First, the National Park Service assumed a more prominent posture in the federal relief archaeological scheme, as the Smithsonian's role diminished. No longer responsible for the initiation and design of archaeological projects, the Institution instead reviewed them for the Works Agency. Moreover, two years passed before the Smithsonian aggressively attempted to influence archaeological policies in the WPA, this done through a consultant appointed within the relief agency. Second, the impact of this new bureaucratic condition on the coalescence of archaeology was considerable. All archaeologic work relief projects, encompassing a wide array of activities, now competed against all other proposals for finite funds; no longer were proposals filtered through the friendly confines of the Smithsonian. Stated differently, a new but neutral body dictated a novel working environment, which affected the political and professional evolution of the discipline. Archaeology's response to these changes would necessitate formulation of further political strategies, strategies designed to enhance the discipline's role within the federal government. The reaction came in two stages: initially, by personnel within the federal government, including the WPA, and then, by the community as a whole.[8]

The organizational infrastructure of the Works Progress Administration determined the nature of the next phase of archaeology's working relationship with the federal government.[9] In general, the results achieved by the operation of this relief agency in its first two years were mixed, in good part due to ambiguities in the initial promulgated standards. The Division of Information and Applications in 1935 developed the following guidelines:

1. Projects should be useful.
2. Projects should require the expenditure of a considerable proportion of the total cost as wages for labor.
3. Projects would be specifically favored that promised an ultimate return to the federal treasury of a considerable proportion of the cost.
4. Projects should be capable of immediate prosecution, so that the money would be spent promptly.
5. Projects should be such as to provide employment to those on the relief rolls.
6. Projects should be located in areas proportionately to the number of workers on relief rolls in those areas.
7. Projects should be such as to facilitate the movement of a maximum number of persons from the relief rolls to work in the shortest possible time.[10]

Once the Works Administration began to accept applications for myriad types of projects, the response was staggering. By September 1935 the agency had received 16,306 petitions, with a price tag in excess of $7 billion. Review of archaeological proposals occurred in this hectic atmosphere. For archaeology, in particular, the review process concentrated on four criteria: low man-year cost, so the funds could be distributed over three years; the ability of relief personnel to perform most of the labor; proximity of the project to populations in need of relief; and the stipulation that single projects in a geographic area not be so labor intensive as to effect a drain on the available pool of labor, thus making other types of relief efforts more difficult to operate. Scientific standards stood noticeably absent from this listing of criteria. No formal mechanism surfaced to assure that quality, professional archaeology resulted from these digs. Here, the Works Administration deferred to the Smithsonian and the "community" at large. The WPA operated under the assumption that archaeology possessed sufficient organizational capacity to police itself.[11]

Part of the original Executive Order that created the Administration imposed an accounting feature later expanded in scope. Order No. 7034 directed the Works Agency "to formulate and administer a system of uniform periodic reports."[12] It should also be understood that the Women's and Professional Division (WPD) for most of the New Deal era administered the archaeology programs.[13] This agency supplied report forms to monitor the many programs, tools not fully suited to archaeology's needs. Carlyle Smith recounted that, early on, supervisors of archaeologic relief work in the South submitted reports on a form designed for use on sewing projects.[14] Fortunately, the Division later created documents appropriate for archaeology. In still other instances, directives from agency administrative managers requested that field directors simply fill in those sections of the forms that seemed applicable.[15] The dearth of such specificity and foresight in this matter aside, it is also apparent that archaeologists' not infrequently expressed dismay about this system stemmed from the absence of any conception of the bureaucratic value of such paperwork. Ultimately, the Works Progress Administration had to collate all the information generated from relief efforts, including attendant expenditures, in order to present an accounting profile to the Government Accounting Office and the appropriate congressional committees.

A related complication that arose centered upon the need for the Works agency to quantify and condense data contained in project reports. How did one quantify scientific material in the form of artifacts? The simplest approach for the accountant focused on how many had been collected. Yet, sheer quantification in and of itself usually held little significance for the archaeologist. But to the bureaucrat, it provided some indication about how efficiently labor had been utilized. Without doubt, the specific nature of the archaeologists' reactions when confronted by this bureaucratic "nightmare" derived from earlier experiences. Previously, and in the broadest sense, they had led a sheltered life within hallowed university confines.

By 1937 the specific problems invoked for archaeological projects peculiar to the grant procedure had become apparent to members of the relief agency, the leaders of the professional academic community, the Smithsonian, and the National Park Service. The new standing of the NPS added another voice among the several member elements of university-trained archaeologists. The growing responsibilities of the Park Service regarding historic and archaeological sites through the 1920s and 1930s demanded inclusion of trained archaeologists on its staff. As more of the relief archaeological work came under its jurisdiction, NPS personnel emerged as another segment of the broader "academic" community that contended for power and monies, but within an economic/administrative environment driven by events essentially beyond their control. A further complication centered on the older tension between the college-based professionals and the widening contingent of archaeologists in government service. Who best to effect judgments bearing on the quality of research? The resolve of this diverse grouping to remedy perceived difficulties eventually resulted in a degree of standardization in reporting and hiring practices. The Park Service, the federal body that had most recently assumed primacy in the archaeological relief and preservation sphere, played a pivotal role in the negotiations that resulted in policy changes. Their participation officially began when A. E. Demeray, Acting Director of the National Park Service, wrote a long letter to Harry Hopkins. It outlined the key problem areas, with the particular object of analysis Operating Procedure No. 04, of which Sections Two (archaeology) and three (paleontology) applied to relief archeological projects. Demeray and the Smithsonian's Frank Setzler also pinpointed an equally troublesome matter. They believed that the Works agency's State Administrators too often failed to consult, seek input, or otherwise communicate through the appropriate channels of the Park Service or the Smithsonian. In fact, not infrequently the National Park Service had to contact such officials and remind them of the provisions of federal legislation that assigned to the Service and the Institution sole jurisdiction in areas of preservation (both historical and archaeological) and scientific investigation (e.g., anthropology, ethnology, paleontology).[16] Demeray stressed that no single institution functioned as a clearinghouse to scrutinize all of the projects. The Park Service and the Smithsonian many times shared that responsibility. Demeray thus recommended, among other things, that a single entity be responsible for approval of projects, issuance of a rigid set of field guidelines, and assuring that progress reports be furnished to the Service. Lastly, he urged that a set of revised instructions be negotiated between representatives of the Park Service, the Smithsonian, and the Works Progress Administration.[17]

Predictably, Demeray's message elicited responses primarily from individuals and agencies within the federal bureaucracy. Leaders of the academic community continued conspicuously silent. Perhaps they failed to assess the larger administrative complexities. Their preoccupation with regional/conceptual areas of study, lack of faith in the practicality of government relief archaeology, and even a disdain for any archaeology associated with government service might have been factors. Irrespective of the exact causation, by July 1938 a revised version of Sections Two and Three of Operating Procedure No. 04, later identified as OP W-18, surfaced. It pronounced restrictive stipulations regarding who could subsequently conduct a dig and specified revised procedural criteria.[18] The Works Progress Administration finally developed, by 1938, a format suited to archaeology and paleontology, prompted by a commitment to create more efficient reporting mechanisms.[19]

Archaeology's relationship with the federal bureaucracy changed in several other aspects. First, after 1938, funding for the Works program came under the purview of Congress: the President and his staff no longer exercised the authority previously held in monopoly (yet another factor outside the control of academic archaeologists that led to further dilution of Smithsonian power). With Works Administration projects now subjected to the annual budget review process, long-lived relief endeavors became a relic of the recent past. Original conceptions about the cooperative relief efforts were altered regarding the responsibilities of the participating state and municipal institutions. Initially, the relief agencies had encouraged state and local authorities to join the undertakings, in part to reduce pressures on the federal treasury. Concurrently, they resolved that local authorities seemed most adept at balancing the twin needs for relief and archaeological preservation. During this relief phase, tighter, more structured, and federally dictated procedural criteria consequently came into play. Moreover, the percentage payment required of state and local relief sponsors increased. This impacted on the small dig, in effect making it a more difficult proposition. Henceforth, only large state institutions or the states themselves usually could afford to finance 25 percent of such endeavors. As a rule, prior to this time most non-federal contributions had ranged from 0 to 10 percent of expenditures. A third way in which archaeology encountered new structures and procedures derived from the appointment of an archaeological consultant to oversee the agency's efforts.

The very fact that the Works Progress Administration, subject to prodding by the Smithsonian and the National Park Service, attempted to erect standards and goals to assure that "scientific" archaeology might result was significant. The introduction of more restrictive, definitive criteria facilitated the movement toward standardization, even if it was initiated by federal agencies. Thus, the story of how the campaign for higher standards evolved merits discussion.

The first step involved the appointment of an archaeological consultant to the WPA: Dr. Vincenzo Petrullo, an anthropologist from the University of Pennsylvania. Complete details about the actual hiring are limited, but Dr. Petrullo's entrance into the WPA came through his contact, Stuart Rice. Petrullo chaired the New Deal's Central Statistical Board, a body designed to keep track of the huge amounts of data contained in incoming government reports. From there, Petrullo migrated into the Women's and Professional Division and discovered relief archaeology. Although Petrullo's background lacked specific training in archaeology, he had extensive experience in Central and North American Indian research and analysis; a complete initiate to archaeology he was not. At this point, he contacted the Smithsonian and immediately created a channel for review of projects. The inclusion of an anthropologist in government service was not unusual. Several anthropologists had involved themselves with New Deal programs for American Indians. But Petrullo stands as the first such professional outside the Park Service and the SI to influence the course of WPA archaeology. Once there, Petrullo attempted to implement some basic anthropological (and therefore, to a large degree, archaeological) tenets and approaches.[20]

The impetus for better standardization in good part resulted from attention directed to two items: the monthly (later quarterly) reports and a standard field manual. Vincenzo Petrullo penned a letter to William Webb in April 1938 in which he requested a copy of the field manual used by the University of Kentucky and the Tennessee Valley Authority. Petrullo planned to distribute several hundred copies of it. He reasoned that if all funded projects were organized in a uniform manner, the methods utilized and the information obtained would be fully comprehensible to the next generation of students.[21]

More extensive scrutiny and control of the relief expeditions also involved greater personal attention by the archaeological consultant and his assistants. For example, Petrullo at times visited the sites and the laboratory facilities.[22] In addition, both project applications and monthly reports now received closer examination. One recorded incident followed a communication by Ellen Woodward, Assistant Administrator in charge of archaeological work, to George H. Goodman, the Kentucky State Administrator of the WPA. Responding to Goodman's previous contact, she suggested that future reports include more detailed scientific data, replete with descriptions of materials and sites, more carefully quantified observations, and any solutions to archaeologic problems revealed by the expedition's experiences. She also inquired about such things as the number of specimens unearthed, the quantity and quality of the sites investigated, the kinds of maps utilized or developed, and the status of the field notes.[23] It must be emphasized that Goodman's diligence did not represent simply an isolated example of the Works agency's initiative in such matters. Non-archaeologic personnel also took an interest in assessing the quality of work.[24] Thus Petrullo's self-spawned activities found support by the officials in the Women's Division as well as the SI. Each action underscored the shift of power from the Smithsonian to other federal bodies like the NPS. Later complaints from professional university archaeologists suggested that the Works Administration lacked commitment to the development of adequate standards. This notion ignores the fact that by this time, the college and museum practitioners, along with the government-based professionals, had begun to maneuver to control more carefully relief archaeology. Petrullo agreed that this responsibility lay with the profession.[25]

One final facet of the changed administrative environment that affected measurably the federal government–archaeological relationship centered upon the National Park Service. Since the turn of the century the Park Service had increasingly become a more important instrument in the historic preservation crusade. Construction and maintenance of the national parks and monuments allotted more power to this agency, initially by means of the 1906 Historic Sites Act, and later through legislation implemented after World War I.

The central role of the Park Service in preservation came from two subsequent actions. First, Roosevelt's Executive Order 6166 of 10 June 1933 increased the geographic area under Service jurisdiction and quadrupled the number of historic sites managed.[26] The action shifted to the NPS responsibilities previously held by the War Department for preservation of battlefields.[27] Conditions and circumstances peculiar to the New Deal period also provided further opportunity for the Service to expand its influence. New positions for professionally trained anthropologists and archaeologists were created. Additionally, since the outmoded 1906 law continued to govern the role of the Park Service, representatives sought more novel, comprehensive legislation.

The Office of the Chief Park Historian and the agency's legal services division drafted a bill more clearly defining NPS preservation responsibilities. They introduced it to Congress in early 1935. The bill passed with little opposition and Roosevelt signed it into law in September 1935. The act allowed for

> the creation under service leadership of a national program for the appropriate preservation, development, and public use of the nation's great historic places.[28]

Archaeological sites likewise came under the newly enhanced powers of the agency.

The Park Service faced two distinct problems as it entered the New Deal era. It lacked sufficient numbers of trained personnel to carry out its tasks effectively. Neither did it possess adequate technicians to oversee specialized operations. The Service's response to this insufficiency involved the creation of the Archaeological Sites Division and the appointment of its first director, A. R. Kelly, who also retained his responsibilities at Ocmulgee. In one sense, initial opposition to this action by the Smithsonian soon disappeared, since Kelly retained his close relationship with Frank Setzler.[29] Nevertheless, the Smithsonian maintained its primacy as the only government agency fully qualified to pass judgment on archaeological matters, despite the fact that the new statute precluded just such activity, thus "officially" reducing the SI's influence. Regardless, a spirit of cooperation, to some extent based on the Setzler–Kelly relationship, prevailed.[30] Some support also came from the Institution through Kelly's proposal to unify all the archaeological work in the southeast under the aegis of one Works Progress Administration project. Initial backing for the idea withered from want of sustained interest and support. It owed nothing to the lack of a co-adjunctive position of the two organizations.[31]

Entrance of the Park Service into the relief program network impacted in several ways. First, although to varying degrees, more university-trained archaeologists gained visibility and influence within three federal bureaucracies. This increased their ability to direct policy formulation on two levels: one, to pursue generally consistent scientific approaches learned in academic preparation; and two, to enhance the needs of their own agency. Simultaneously, both the Works Administration and the Park Service employed growing numbers of trained personnel, who themselves proved useful in meeting and enforcing further adequate scientific standards. Third, a more sympathetic voice had been centrally placed within the WPA decision-making apparatus: Vincenzo Petrullo, a trained anthropologist familiar with archaeological scientific methodology.

One example demonstrates how this system worked to the interest of the community and addressed its concern over standardization of field and laboratory methodology. Petrullo early on alerted Matthew Stirling in the Bureau of American Ethnology about the imminent, intensified emphasis by the WPA on the conduct of archaeological expeditions. Stirling quickly responded by writing a letter to Ellen Woodward, in which he commended Petrullo for the courtesy extended. He also assured her that the Works Progress Administration would continue to receive support from the Smithsonian.[32] Without question, the Smithsonian still remained capable of exercising influence, even though the Park Service largely fulfilled the central function previously held by the Institution, that of assessing the work of project applications. Each proposal was judged based upon multiple criteria, including the project's academic merit, the number and caliber of supervisors specified, monetary costs relative to the scope of the excavation, intended use of improved investigative techniques, and arrangements for publication of data. Still, when the Works agency sought personnel for supervisory positions, they deferred to the Smithsonian. Usually, state relief headquarters contacted the Smithsonian directly for a list of eligible professionals. Those recommended usually had association with prominent academic institutions and museums.[33]

By May 1938 the new level of cooperation shared by the Park Service and the Smithsonian Institution assumed a more formal posture. Secretary Abbot proposed to Secretary of the Interior Harold Ickes that the two organizations pool their efforts to undertake work on the Historic Sites Survey. The Institution's staff labored within a framework designated by the National Park Service to maximize utilization of the abilities of the Smithsonian work force. The Park Service, which welcomed the offer, paid for any costs incurred in travel and preparation of reports. Additional details of the government venture became the responsibility of a committee manned by familiar names: Frank Setzler, Matthew Stirling, Arthur Kelly, and the Park Service's acting Assistant Director, Ronald Lee.[34] The inter-agency liaison strengthened its mission in May 1939 when Secretary Ickes invited Setzler to become a member of the Advisory Board of the National Park Service. By the end of the decade, professional academic archaeologists and anthropologists permeated the highest levels of government decision-making.[35]

In concluding this chapter, two sets of observations merit further comment. The first set of observations encompasses four points One, the new or altered relationships spurred by the WPA relief programs increasingly assimilated greater numbers of academically trained professional archaeologists into the decision-making apparatus, either on a permanent basis or as temporary supervisory staff. There, these individuals influenced further hiring practices, methods of fieldwork, publication strategies, and other approaches and procedures consistent with their professional training and background. Two, the campaign to erect a more standardized reporting mechanism dovetailed with efforts pursued independently by university archaeologists. Three, a more composite academic discipline (both government and university) in particular benefited from the availability of massive amounts of new information, the continued dissemination of improved field methods, increased work opportunities for graduate students, and the organization of new archaeological conferences, all products of federally funded relief archaeology. Four, with New Deal–fostered experiences and lessons on hand, an elite cadre constituting the discipline's academic leadership began to understand the need for their own organizations to influence directly the federal bureaucratic structures.

Pertinent to the place occupied in history by the Smithsonian relative to New Deal archaeology, several points require comment. The role of the Smithsonian changed greatly from the time of the Civil Works Administration due to circumstances that it did not control. This fact aside, some post–New Deal communications criticized the agency, alleging that the Smithsonian failed to ensure sufficiently high standards early on in the life of the Works Progress Administration.[36] Precisely, only the Works Agency possessed the final authority to approve or reject project proposals. The Smithsonian's role instead assessed a suggested project's scientific value, the reasonableness of specific costs, and the expertise of the designated leadership. If a proposal required alteration, as determined by the Smithsonian, WPA headquarters returned the paperwork to the initiator for modification. Relief proposals moved upward from local agencies and sponsoring institutions; Washington determined the manner and extent of emergency aid. The primary concern of the Works Progress Administration always focused on the needs of the public. In passing judgment, one must remain cognizant that the Smithsonian had neither the staff nor the money to monitor closely programs that were spread all over the country. For that task, the Works Administration ultimately depended on the man in charge of the project. The Smithsonian had capably supervised projects during the Civil Works Administration. The WPA expected archaeologists outside the Institution to do the same, since the final responsibility for the quality of the expedition rested with the personnel on location. If disparities of quality existed between the projects, such disparities testified to the persistent presence of variations or discontinuity in the application of methods, training received by personnel, and guidelines followed within the community. If the archaeologists thought that the development of higher standards was the sole responsibility of the inner workings of a bureaucratic relief apparatus, then they failed to perceive their own professional shortcomings.

Chapter 5

Relief Archaeology: Practice and Political Consolidation

An analysis here of the archaeological expeditions of the Works Progress Administration, chosen as representative of the significant shifts that ultimately led to greater national and regional consolidation and political maturation, reflect other key aspects of the discipline as well. The scope and length of the digs, combined with the WPA's continued approach to publicity, carried the message of the importance of archaeological work and its relief benefits. Press releases carried the message to the public with headlines like, "Archaeological Projects Seek Light On Early Americans."[1] The result: the terms "scientific" and "archaeology" were increasingly deemed synonymous. The next generation of leaders, Gordon Willey, Jesse Jennings, William Haag, James Griffin, and James Ford, to name a few, continued to find career opportunities while maintaining the basic role of fieldwork in archaeologic investigation. This occurred principally through the Works Progress Administration, as it broadened extensively relief efforts. A host of other alphabet agencies also assumed a role. The increased centralization of funding and relief through the aegis of the federal government also prompted the field's leadership to seek political solutions, so as to influence the direction of programs that affected their interests. Equally important, the debate surrounding questions as to what constituted proper archaeology sharpened ideas about the nature of training, laboratory analyses, and the necessity for publishing. Therefore, the relief programs facilitated a freer flow of ideas, forced decisions regarding methodology and syntax, and produced much more data than ever before. This, in turn, prompted revision of chronological sequences and macro cultural concepts. In the effort to address such matters, academic archaeologists continued to consolidate politically in order to gain increased control over their specialty, in part by defining the practice of archaeology as inclusive of a political component. Programs that incorporated such notions at the national level constitute this discussion. The previous period witnessed mostly government-based specialists assuming the initiative, with the remainder of the academic community soon joining the fray. While full cooperation among disciplinary elements both inside and outside the government was initially halting, ultimately both worked toward twin common goals: to gain predominant representation for professional archaeology's conceptions and to assure the inclusion of professionals in the decision-making process. The digs themselves provided the mechanism by which such consolidation strategies were initiated.

Expeditions in Kentucky and Tennessee and the other Deep South states produced significant new documentation. Information from the Atlantic Coast and Great Lakes states defined cultural relationships more specifically, thereby bringing a greater understanding of these group's history. Considerable effort along similar lines occurred in the Southwest and Great Plains. All these endeavors produced results based upon different levels of disciplinary consolidation and cooperation. This made all the more fascinating the nature of the institutional growth of this field. The sheer size of the United States and the multiplicity of institutions at the state and local levels generated a multilevel evolution. The historical matrix of archaeology witnessed simultaneous incorporation of its many elements on different levels at dissimilar rates through time. Perhaps the single best example involved the regional conference, which flourished in this period.[2]

The conference offered considerable opportunity for effective communication. It was ideally suited to a profession still relatively small in size. Meetings, never numbering more than thirty to forty individuals, enabled a majority of the "elite" archaeologists to gather and exchange information in a single forum. Moreover, new and younger members were neither isolated nor discouraged. They produced similar, and at times, even higher caliber work than their senior colleagues, thus rightfully demanding the parity of status they felt was due them. In this sense, the New Deal relief expeditions became a vehicle for the invigoration of the archaeological discipline. Initiates and established members alike laid out the latest artifacts, collaborated face to face, and discussed the implications of their research, with immediate critique and feedback available. Relief efforts acted as another means to reinforce the ideal of a professional community. Due to the nature of the programs, the archaeological participants came from all over a region and throughout the country, a process that might have occurred in other disciplines only in more prosperous times. The annual meetings of the Society for American Archaeology accelerated this process, but did not give birth to it. Finally, the huge amount of material unearthed created a momentum of its own. Something had to be done about assimilating the tons of new material into classification and temporal systems.[3]

True, not all of the work done under the auspices of the Works Progress Administration reflected the standards of academic and government leaders. Nevertheless, in general all elements of the discipline, at every level, were affected favorably: government or private, college or museum, and state and local institutions. Writing in 1961, William G. Haag summed up the cumulative effects:

> The last 25 years have produced a rather revolutionary change in our knowledge of Eastern archaeology. Not only has the gross content of our knowledge increased but the theoretical orientation as well has undergone a marked change.[4]

By far the most well known and widely publicized New Deal archaeological expeditions took place in the South. The vast number of popular and academic articles about WPA archaeology also focused on this region. Since many postwar leaders spent considerable time there, anecdotal chronicles and research published by them often reflected on those 1930's experiences. Discussion of Depression-era archaeology generally evoked images of the huge and spectacular digs in the Tennessee River Valley. The area also constituted the last great link in a system that led to a total cultural conceptualization of the eastern United States.

Works Progress relief efforts in Louisiana revolved around one man, James A. Ford. Since more often than not a single individual largely determined the nature and scope of programs in one state, this was not an anomalous circumstance. Ford's knowledge, gained in previous New Deal work, coupled with the personal and professional relationships he had cultivated with other members of the archaeological community, enabled him to orchestrate the most efficient relief operation in the South. Field management, laboratory processing, and the rate of publishing proved particularly effective. Most notably, under Ford, exercises in laboratory and fieldwork served to train virtually every leader of the next generation.[5]

Institution-building also marked this state's work. The Louisiana Works Progress archaeologic projects further established the hegemony of Louisiana

State University as the primary center for training new disciples. In turn, this strengthened the position of large-scale, university-based archaeology over smaller regional units.

Pursuant to the subject of the strong personality and its impact on the evolution of 1930's archaeology, it is equally important to understand the fundamental ambiguity that existed with regard to enforcement of standards within the federal relief programs. The Women's and Professional Division depended on the appointed archaeologists representing the still-emergent community to self-monitor behavior, overcome strong regional identification, and promote better management practices.[6] This posture had occurred in other instances during the Roosevelt presidency. For example, the National Recovery Administration encouraged the business community to use the Blue Eagle to erect its own standards and codes rather than have them dictated by the government.[7] The Works Progress Administration, because of the breadth of its jurisdiction, relied on the multitude of sponsors to do the same. In this manner, Hopkins wanted the archaeological community itself to insure the practice of "good" archaeology. Because of Ford's consummate abilities, reliance on this philosophy served well the interests of both archaeology and the Works agency.

The chronicle in Louisiana began with Ford as a graduate student at the University of Michigan, when the Works Progress Administration programs commenced. His education at the Ann Arbor campus introduced him to the most current research methods and leading experts in the field. Close associations with many other students occurred as well. When he returned to Louisiana, where he had received his undergraduate training, he assembled around him a quality group of bright and energetic individuals.[8] They included Preston Holder, Carlyle Smith, Gordon Willey, George I. Quimby, Jr., Arden R. King, William T. Mulloy, Robert S. Neitzel, and Edwin B. Doran, Jr.[9] Ford had met many of these men either in college or as fellow participants in his earlier New Deal work in Marksville and Georgia. The university-trained men came from throughout the country. For example, Willey had attended the University of Arizona and apprenticed at Macon before he arrived in Louisiana. Later, he pursued his doctoral studies at Columbia University. Neitzel, a graduate from Nebraska, had previously participated in fieldwork in Tennessee. Doran studied at Louisiana State University, Mulloy and King attended the University of Utah, Holder arrived from Columbia, and Quimby from Michigan.[10]

Ford's Louisiana expeditions divided time between fieldwork and extensive laboratory analysis. His carefully selected sites, not inordinately large and employing usually thirty to forty- five men, funneled an abundance of material to the laboratory facilities. Louisiana State University, the primary sponsor, functioned as the central administrative body for relief work. The continued presence of the Works Progress Administration on campus spurred the establishment of the School of Geology, a composite structure that incorporated the Departments of Geography, Anthropology, Geology, and Petroleum Engineering. The eventual impact of the increased visibility and status of archaeology manifested itself in the appointment of William G. Haag to the first chair of archaeology at LSU in 1952.[11]

Through previous professional archaeologic work dating from Winslow Walker's work in Troyville and subsequent Smithsonian efforts, the New Deal advanced large-scale, academic archaeology into the state.[12] Convinced about the viability of relief projects in Louisiana, Ford prepared and submitted a plan in 1938. For this, he secured the aid of Gordon Willey to organize a staff of supervisors. Approval of the project brought funding in excess of $120,000, with much of the money allocated for laboratory analysis.[13] The lab, originally located in New Orleans rather than Baton Rouge because of lower rental rates, later moved to the campus at Louisiana State.[14]

Operation of the laboratory unit provided a prime example of how archaeologists had to improvise and then agree on common practice, thus adopting the procedural, managerial approach to archaeologic analysis in light of new needs and conditions. Importantly, this phenomenon was aided by the input and comment of representatives from several institutions all around the country. The laboratory organization facilitated careful manipulation of that environment, with analytical tasks compartmentalized into discrete labor units. Incoming material progressed through stages that included initial analysis, statistical evaluation, structural "engineering", use of photography, application of dendrochronological technique, and cataloging and recordkeeping. Other necessary sections provided carpentry and administrative/clerical services. In short, sharply delineated laboratory procedures became systematized.[15]

The reasons why Ford and other archaeologists opted for this approach related directly to the unique circumstances of relief archaeology. First, the time strictures that applied to WPA hiring policies precluded long-term employment of personnel. The resultant constant turnover of workers meant that insufficient time existed to train them in all facets of laboratory work. In response, the archaeologists developed clear-cut instructions to guide each person in but one or two specialty tasks. Another prime consideration stemmed from the fact that relief workers had no formal training in anthropology. They could not be expected to comprehend the totality of laboratory analysis. Finally, the material from the field inundated the laboratories. Simply stated, the size of the expeditions created new pressures for rapid processing of data.

Implementation of such a comprehensive and structured laboratory organization resulted in a tremendous amount of information being made available to many people. The precedent firmly established by Ford in Louisiana—thorough, rapidly paced, and highly productive laboratory analysis—became an integral component of the evolving research technique.

Fieldwork in Louisiana required an even larger contribution of time and labor. Road building, in addition to preliminary survey and excavation, often accompanied the recovery process. Ford targeted his excavations to collect ceramic remains, as did other eastern- and northern-trained archaeologists. Pottery and ceramic shards embodied the single key to creating a classification system and a chronology for prehistory in the East.

Work began in October 1938 and continued until June 1941. Exploration at the LaSalle–Paris site, and specifically, the Crooks Mound, helped to reveal early Marksville specimens of points, tools, burials, and shards. Other related work exposed Coles Creek Horizon.[16] The other major aspect of these expeditions took place later, in West Baton Rouge and Avoyelles, where the Greenhouse Mound produced indications of early Tchefuncte.[17] The work in this culture helped to establish the Tchefuncte, an older culture that predated Marksville. An older Woodland culture, distinctive by its form of pottery, derived from the even older Archaic.

The results from these programs prompted a revision of Ford's own ideas on chronological epochs as he expanded his understanding to include the new period discovered. Another important feature of the Louisiana operations centered on the extensive publication record of these young archaeologists. Almost as soon as workers processed the material, it found its way into print. These documents, though not as encyclopedic as the Webb reports for the TVA, contained information of a high quality.[18]

One press release for the Louisiana expeditions was representative of how the WPA sought to equate work relief with quality research. The headline read "IMPORTANT ARCHAEOLOGICAL FINDS IN LOUISIANA." A description of the project followed, with the last paragraphs reading:

The men and women assigned to the archaeological project represent varied trades and professions. The excavating is done by laborers, who are painstakingly schooled in the process of taking delicate artifacts from the earth without damage. Employed in project headquarters are clerks, statisticians, draftsmen, artists and photographers.

Trained archaeologists, of course, then direct the work of studying the habits of pre-historic man.[19]

As might be expected, in Kentucky, William Webb dominated the scene. The course of the Kentucky Works Progress Administration and the Tennessee Valley Authority investigations, under the leadership of Webb, completed the transformation of Kentucky archaeology. Concurrently, that experience catapulted Webb into national prominence. With his usual strong control, Webb dictated where and how most digging would occur. Not since the nineteenth century had one individual exercised so thorough control of an area. In terms of accumulating information, the knowledge gained from Webb-led digs proved pivotal in the understanding of prehistoric Kentucky. The incredible amounts of data and the vast number of publications aided later syntheses, such as J. B. Griffin's work on the Fort Ancient Culture as well as the wider eastern area.[20] Webb also functioned as an institution builder. Archaeology became established at the University of Kentucky as something more than a part-time avocation. Last of all, the Major helped assure that a group of southeastern archaeologists received training for later years of work and research.

Webb's approaches to the Kentucky projects, both through the Works Progress Administration and the Tennessee Valley Authority, reflected his previous experiences in the TVA. From similar experiences with labor-intensive work, which had uncovered sites in Kentucky, Webb already understood that a well-managed, large operation could reap appreciable dividends. With a focus on large-scale operations, Webb then targeted digs only if he was assured of ample supplies of available labor. He broke down selected areas into distinct sectors after survey and sample excavations had been completed.

Archaeologically, Kentucky borders and encompasses several prehistoric Indian cultures, along with several natural throughways and pathways. As such, several cultural epochs appear in this state's prehistory, each varying according to place and time. Webb attacked Kentucky in an attempt to fill gaps in information about time eras and cultural type. The Major based his approach on two criteria: the need for work relief and the desire to fortify his own interpretive apparatus.[21] He directed superficial excavation of many sites in each of the designated areas. One or two were then singled out for particular examination. His trusted supervisors included C. T. R. Bohannan, Ralph D. Brown, J. L. Buckner, John Cotter, John B. Elliott, James C. Greenacre, William G. Haag, Claude Johnston, Carl F. Miller, Albert Spaulding, and David Stout.[22] The sheer number of these individuals indicated not only that he delegated authority, but also the size of the areas receiving attention. Haag, the person closest to Webb and entrusted with important responsibilities, shared the later rewards of publication. The other men, trained in fieldwork, had academic backgrounds conducive to this type of work.[23]

In his research, Webb investigated the classifications of early hunter or Archaic, Adena, Woodland, and Mississippian cultures. The Woodland and Mississippi included areas in Kentucky and beyond. Archaeologists, at that point, understood more about the Archaic and Adena regions outside the state.[24] The Works Progress Administration archaeological relief ventures expanded the knowledge of these two cultures considerably. The fact that the projects encompassed ten counties reveals the scope of the investigations. Keep in mind another salient fact: Webb successfully garnered large sums to finance these operations.[25] In 1937, the University accepted more than $75,000 in support funds, an amount that permitted the organization of endeavors that would uncover enormous numbers of artifacts. In 1940 alone, the Kentucky work exhumed 1,750 skeletons and 48,000 artifacts, and took 3,700 photographs; and this represented one of the smaller outputs of data.[26]

In western Kentucky, Webb sought evidence of the Archaic. The term, invented by William Ritchie, referred to sites in New York and described a culture that produced no pottery and was pre-agricultural.[27] Webb also found other cultures, like the Mississippi.[28] In this instance, the structure of the mounds, in terms of size and complexity, made the investigation costly; use of WPA labor and money made it possible. Webb led excavations at other, equally complex sites that required immense numbers of man hours.[29] Except for the unique circumstances, which made relief aid available, exploration of these sites might have been precluded for years.[30]

William Haag supervised the laboratory portion of these Kentucky projects. The lab processed, cleaned, restored, repaired, and catalogued over 55,000 artifacts that ranged from early to late Archaic. The exposure of the depth of the culture in Kentucky helped to clarify the conception of the pre-Woodland epoch in the East.[31]

The 1930's efforts made identification of the Adena culture in Kentucky more clear. Marked by conical burial mounds rich in both bodies and possessions, the use of cremation constituted another prominent feature. This culture interested archaeologists greatly, for they thought it might represent a culture distinct from Fort Ancient and Hopewell, although initially Webb hesitated to categorize it as such.[32] Nonetheless, the Adena work completed here helped to establish the breadth of the culture and substantially added to the list of distinguishable traits. Webb eventually produced a table of four Hopewellian manifestations, which identified 121 traits, including forms of house construction.

Ultimately, William Webb did view the Adena as a distinct cultural phase. James Griffin, though, interpreted the Adena as a variation and a continuation of Ohio Hopewell.[33] He envisioned it as representative of both early and middle Woodland. The divergent perspectives of the two men revealed a significant point. Webb held a more narrowly defined base of knowledge, one derived from his Kentucky work. Accordingly, he failed to foresee a larger cultural complex beyond the state. Griffin adopted a broader perspective, one that encompassed the entire eastern half of the continent. In reality, no case better illustrated a fundamental truth: both men reflected the transition underway in American archaeology. Webb, the consummate manager and politician, demonstrated that archaeologists in the future would have to cultivate political acumen. He lacked, however, the ability to conceptualize on a grand scale. Griffin, on the other hand, had been specifically trained as a ceramist and a theorist. Beginning in this period, the specialist would become more prominent in archaeology.[34]

Aside from the broad interpretive differences Griffin and Webb attributed to the data, the considerable cultural remains unearthed proved noteworthy in other ways. Substantial ceramic evidence from the C & O Mounds in 1939 established western occupation. Further proof of Adena culture became available with the discovery of extensive burials, cremations, and ceramic pots in the Mount Horeb Earthworks and the Morgan Stone Mound.[35] Adena culture extended even into eastern Kentucky.[36] Rock shelters there showed a people who used corn, sunflower seeds, squash, and other items associated with the Adena. Irrespective of this evidence, Webb failed to identify a connection between them. Archaeologist Douglas Schwartz later noted that Webb was remiss in not addressing important questions about the relationship between the two groups. Nor did Webb posit queries about which group initiated domestication. Webb's penchant for demanding clear-cut delineation can be measured in another way. The Major adopted the Midwestern Taxonomic System, popular at the time, to fit the information he had about the time/culture epochs in Kentucky. In short, he either ignored certain types of information or dismissed the data as anomalous. Griffin's interpretive skills were needed to perceive the discrepancies that represented the transition of one culture to another.[37]

The Webb expeditions also examined Fort Ancient, another Ohio Valley culture. However, he devoted far less time and effort to these digs. Fort Ancient lacked burial mounds but did have pottery and shell work, some evidence of agriculture, and large villages. Still, Webb found it far less engaging than Adena and Archaic. Another possible reason for his decided lack of interest centered around the potential confrontation: Fort Ancient constituted a principal component of James Griffin's domain. Any extensive involvement here by Webb might have evoked interpretive conflict with a close colleague who honed the cutting edge of disciplinary synthesis.[38]

The last major group of inquiry for Webb in Kentucky involved the Mississippian culture. Centered in the more expansive rivers in western Kentucky, large villages enclosed by stockades, planned communities with temple mounds, and a domesticated agricultural economy marked this society. Excavation here, which utilized Civilian Conservation Corps labor, continued until the outbreak of World War II.[39]

In offering an assessment of Webb, one must remember that he changed the expression of Kentucky archaeology. It evolved from a pursuit that enjoyed an amateur status to an academic profession based in the university. Even though Webb spent relatively little time in Kentucky, engaged as he was in "emergency" work with the Tennessee Valley Authority, his capable supervisors ran the Kentucky-based projects. That fact symbolized one of his great strengths: the ability to assemble a competent and loyal cadre of subordinates. These extensive digs produced more information on the Archaic and Adena than ever before. His leadership resulted in the continual release of important data that allowed other professionals, notably Fay-Cooper Cole, James Griffin, Gordon Willey, and James Ford, to offer more elaborate and ambitious interpretations of eastern archaeology. This rapid development occurred in less than fifteen years, in an environment devoid of the strong presence of a museum or a state historical or archaeological society. Instead, it focused on one institution: the University of Kentucky. The seeds for this transformation had been planted in the late 1920s, with Webb and Funkhouser's self-financed summer digs. Also, the expansion of standardized field methods and their application followed. The inclusion of supervisors drawn from institutions around the country aided in this.[40] In addition, new life was brought to the Southeastern Archaeological Conferences attended by most of Webb's assistants. These affairs allowed the men to interact on a personal and professional level and to speculate with other eastern archaeologists about anticipated research problems.[41]

Tennessee archaeology offered an interesting contrast to that of Kentucky. The role of personality continued to be significant, as in Kentucky and Louisiana. However, Tennessee's T. M. N. Lewis lacked the equivalent dynamism and capabilities of Ford and Webb. Nevertheless, he profoundly affected the course of 1930's Tennessee archaeology.

Tennessee during the 1930s followed a pattern of development initially similar to Kentucky's. The University of Tennessee became the focal point of archaeological programs. Acting from a suggestion by William Webb, the university president, James D. Hoskins, in September 1934 moved to create the Division of Anthropology within the Department of History, with Lewis as director. Lewis occupied a status equivalent to Webb's: a talented amateur who had sustained research through personal means. From 1934 on, if applications received approval, the state could carry out federally funded surveys and expeditions. Complications arose when Lewis proved himself basically incapable of organizing and managing a large-scale archaeological project. Problems quickly emerged that revolved around failures to process material and Lewis's sometimes vitriolic personality.[42]

A lack of adequate planning by Lewis evoked the first of a seemingly endless number of confrontations that plagued him. The original Works Agency appropriation proposal contained a flaw: Lewis had not provided for laboratory work. Accordingly, lab processing on this first project did not begin until four years later, in 1938. Such deficiencies, plus a dismal publication record, brought Lewis into conflict with the Works Progress Administration. His reactive personality only worsened the situation. In Kentucky, Georgia, and Louisiana, by way of contrast, personnel processed material obtained from their digs in as little as two weeks. That data also found its way into at least a preliminary form of publication with relative speed.[43] Regrettably for Lewis, superior standards in use elsewhere (which, incidentally, were the product not of the Works Progress Administration but of certain portions of the southern relief academic community) meant the program in Tennessee did not measure up well. The Works Progress Administration archaeological consultants, Vincenzo Petrullo and his later replacement Stella L. Deignan, consistently put pressure on Lewis to improve his performance.[44] The concept and practice of rapid and prolific publication had become an accepted part of the emergent archaeological research norm. The precedents set by Webb and Ford strongly influenced this process.

Rather than accept the need to alter his methods, Lewis consistently attacked those whom he believed opposed him. this was not a sound tactic, as it alienated other professionals, William Webb included.[45] At a crucial historical moment, when a significant degree of unification in the southern archaeological community seemed imminent, Lewis' actions ran counter to that pattern. Desirous of support, he sought assurances from others in the academic community. At his invitation, Carl Guthe, a longstanding friend, came to Tennessee and inspected the digs. He pronounced them of sufficient quality.[46] Yet, the publications and processing lagged.[47]

Lewis used monthly reports—long, detailed chronicles of research activities—as his primary vehicle for the dissemination of information. A scant circulation base limited their visibility. Lewis and his students uncovered village sites, mounds, and burials in nineteen counties. Many sites, of a multi-occupational nature, were associated with Woodland and Mississippian. The digs produced considerable numbers of pottery artifacts with a limestone temper, cord-marked, plain, and stamped, with some Archaic included.[48] Lewis submitted the final report on Chickamauga in June 1941. Later published in a preliminary form, it consisted of information and descriptions of thirteen major sites judged the most important.

Lewis's approach to Tennessee archaeological investigation differed markedly from William Webb's.[49] Since a majority of Lewis's investigations were determined by the Tennessee Valley Authority, he exercised little choice in where to explore. The designated locations for the proposed dams precluded such decision-making. Thus, Lewis could not devote as much time to Works Progress Administration endeavors as he might have wished.

The remainder of relief work in Tennessee occurred around Chickamauga, where Jesse Jennings supervised the WPA labor.[50] After an initial excavation of 70 sites, another 220 subsequently received attention.[51] Again, Lewis generated animosity when he and Webb butted heads over publication strategies. Lewis wanted to author the research findings. Webb believed this to be his privilege since he had penned all the other reports. Lewis then opted to write a short article, replete with photographs. Webb deemed unprofessional Lewis's use of information borrowed from another. In a series of heated letters, they chastised each other.[52] Ultimately, neither Lewis nor Webb published any material on the site; only the monthly report was submitted. Because of the preoccupation with wartime activities, the Tennessee Valley Authority never printed it.[53]

A peculiar personal approach to administrative duties by Lewis caused additional consternation. More turmoil ensued. Lewis guarded against any perceived intrusions into Tennessee archaeology. He even adopted a paternalistic attitude toward the Authority work. This posture constituted yet another source of conflict with Webb. When Major Webb proposed further investigations into the Kentucky Basin, where Lewis had been working, Lewis immediately became aroused.[54] Eventually, archaeologists outside the area, personnel within the WPA, and the Tennessee Valley Authority were embroiled in this strife. In short, by temperament Lewis seemed ill-suited to participate in cooperative endeavors. Nor did be communicate a respectful opinion of others. For instance, prior to the falling-out with Webb, Lewis in 1936 wrote disparagingly about several archaeologists who had proven their worth in other excavations. He asserted that Jesse Jennings gave scant attention to field notes. Neitzel purportedly treated artifacts like pieces of coal. He added that the University of Chicago inculcated in its students an attitude toward handling artifacts suitable for amateur collectors but not professional archaeologists.[55]

Such verbiage, which typified Lewis's communications, reflected a vindictiveness and critical mindset that was wholly inappropriate. More likely than not, Lewis's behavior drew from the disdain he believed others held for him. This aside, Lewis simply did not sustain an environment conducive to the generation of mutual respect. For him, either you stood as a loyal supporter or as an enemy. To individuals like Jennings or Neitzel, both of whom had been delegated a great deal of latitude and trust in previous circumstances, Lewis must have seemed an enigma. Other correspondence indicated he fared no better in relations with the hired help.[56]

Another dimension to the Tennessee story transcends these seemingly pedantic controversies and is instead related to the essence of the emergent professionalization movement. The dissension that surfaced involved more than manifestations of ego response. True, there seems little doubt that Lewis possessed an exaggerated opinion of himself, one not often shared by others. Intolerance of divergent beliefs relative to the operation of Works Progress Administration projects prompted Lewis to direct vituperation at others. Lewis rebelled in response to the standards being forced upon him by the WPA. Still other observations hold here. First, the WPA, not the archaeological community in general, attempted to assure implementation and adherence to norms. Equally significant, Lewis exhibited a strong advocacy for "state" archaeology. Simply put, his largely anachronistic views identified with the "old" archaeology, practices characterized by state hegemony and the presence of a limited national institutional structure. (Paradoxically, Lewis is credited with the "seed" idea of what eventually became the Society for American Archaeology). More than personal animosities were, of course, at work. The massive expeditions fostered an environment for academic archaeologists to work together, exact conclusions about their fieldwork, and give thought to matters of importance to the entire discipline. Ultimately, these operations promoted further political activity focused upon improved control of government operations and related decision-making processes. Interestingly, at one and the same time, this saga revealed insights about the underbelly of the discipline, the strength of past traditions and approaches, and the fact that archaeology was confronted by rapidly changing events over which it often could exert little control.[57]

A remaining facet of the Lewis episode centered upon the problem of standards. Illustrated here is an issue of historical importance. Briefly, one of the major criticisms of Works Administration–sponsored archaeology appeared in the 1940s. It specified that this federal agency exerted little effort to insure that professional quality archaeology resulted.[58] Reasoned insight about both matters can be extracted from further analysis of the Lewis story. The standards by which the WPA judged Lewis did not originate within the agency. Political considerations aside, the Works agency utilized guidelines that emanated from the field itself. it is useful to recall, also, that neither the Smithsonian nor the National Park Service were in a position to promulgate guidelines. They exercised, at best, indirect control over the actual work on a site, and sometimes no control at all. Rather, the promulgated guidelines that were advocated by professionals, both inside and outside the government, generally mirrored the evolving paradigm of university training programs. The fact that a universally accepted consensus had not yet formed attested to the still fragmentary status of the discipline. Ideally, these precepts would have come from a national source, either the National Research Council or the Society for American Archaeology, for example. Instead, experiences gained in the field by the likes of Webb, Ford, even Stirling, Setzler, and Kelly, often under the most pressing circumstances, prompted the creation of "on the spot" criteria. In sum, the nascent practices associated with academic relief archaeology and its recent accomplishments reflected larger changes in the discipline, and involved training and publication. These became the standards that Petrullo, and, in a lesser sense, Deignan, emphasized, thereby conferring official legitimacy on their field techniques. Lewis then was deficient by way of his inability to mimic men like Webb. In conclusion, the charge of malfeasance leveled at the Works Progress Administration lacked substance. Standards, to the degree they existed, came from the community and originated in the earliest days of relief in the CWA.[59]

Additional comments about the role of personality in the history of American archaeology seem appropriate. The cases of Webb and Ford are appropriately exemplary. Webb, despite all of the tremendous contributions he made to 1930's relief archaeology was, like Lewis, something of an anomaly, but for a very different reason. Educated as a physicist, he worked as neither the archetypal professional nor as the amateur archaeologist. A unique individual, Webb eludes easy categorization. The most representative transition figure of the region might be Ford. Originally educated in the South, he had migrated northward for professional archaeologic training. He then returned to the South, espousing the most recently developed ideas and scientific procedures. His attitudes and fieldwork embodied most of the features of the archaeologic procedural, or managerial, paradigm: the ability to erect national and crosscultural interpretations and a working knowledge of the latest laboratory techniques and field methods. Ford, consequently, facilitated the dissemination of technical information. He also viewed ceramics as the key artifacts in the creation of classification and temporal systems. Quiet and unassuming in comparison to the more outgoing Webb, Ford and the other younger archaeologists, Haag, Jennings, Willey, and Griffin among them, worked to transform the discipline, even in the face of resistance from within.[60]

The Alabama New Deal archaeology programs shared many of the characteristics of most of the other southern states: strong personalities, concern for state archaeological rights, and the generation of massive amounts of data. None of the unpleasantness surrounding the Lewis episodes occurred, however. All but a very small portion of the work in this state came under the auspices of the TVA projects. The advanced level of cooperation placed this state ahead of several others. This stemmed in part from the fact that a strong, multifaceted institutional network existed. The Alabama Anthropological Society and the Birmingham Anthropological Society were carrying on their own research by the early 1930s.[61] Birmingham Southern College had sponsored digs during the summers of this decade's first years; even Carl Guthe visited as a guest instructor in 1934.[62] The Alabama Museum of Natural History stood as the major academic force in the state, not its affiliate, the Department of Anthropology at the University of Alabama.

The geologist Walter Jones had created an infrastructure throughout the state during his years with the museum in the absence of any archaeological tradition at the major state universities. Jones, likewise, had established contacts throughout the South, as in the case of the Society for Georgia Archaeology. With a strong support staff, headed by the DeJarnette brothers, and loyal and capable supervisors, Jones and his followers believed strongly in expanding Alabama archaeology, albeit in a limited way. As with Webb, Jones and the DeJarnettes wished to develop archaeology within a state framework. Since Jones had other responsibilities within the Museum, the direct supervisorial duties of the Alabama portion of the Tennessee Valley Authority fell to David DeJarnette. Like Haag, Webb, and Ford, he possessed both a marked ability to oversee large operations and a strength of personality that bordered on obstinance; it served him well, nevertheless. The Alabama operations effectively improved the status of archaeology, first through the Museum, then later within the university system.[63]

Pickwick Basin, Guntersville Basin, Wheeler Basin, and surrounding counties comprised the major target areas of inquiry, coupled with surveys of twelve additional counties, all late in the Depression era.[64] Work began less than a year after formulation of the WPA and continued into mid-1942. The investigations heavily explored the northern area of Alabama. Previously unknown aspects of the Tennessee Valley prehistory came to light. Workers uncovered dozens of village sites, hundreds of mounds and caves, thousands of burials, and hundreds of thousands of artifacts.[65] Laboratory processing became the responsibility of the Museum, while the Bureau of American Ethnology and the Alabama Museum assumed publication chores.[66]

David DeJarnette stood as the central figure in the operation of these digs. His organization and managerial skills contributed to Alabama archaeology in a manner similar to the contributions of Ford in Louisiana. Graduated from the University of Alabama, DeJarnette had moved to the Museum, where he first became assistant and then head curator. He also received training in field methods at the University of Chicago in 1932. Later, DeJarnette had worked under William Webb during the Civil Works Administration and Federal Emergency Relief Administration expeditions in Wheeler Basin in 1933–34.[67] His supervisors constituted a cross-section of academically prepared archaeologists, a pattern repeated in other states. They included Harold Anderson, H. Summerfield Day, James R. Foster, Theodore L. Hohansen, Wayne W. Kraxberger, Carl F. Miller, Julie Addock, Harold H. Dahms, Marion L. Dunlevy, and two physical anthropologists, Charles E. Snow and Marshall T. Neuman. Addock, one of the few university-educated women involved in New Deal archaeology, worked in the laboratory (most of the other women involved in archaeological fieldwork worked as unskilled laborers).[68]

The reasons behind the decision to hire Addock are extremely interesting. In a letter to Webb, DeJarnette indicated that "Miss Weber and Dr. Petrullo" exerted pressure to "sponsor an excavation using Negro women." He added, "A project using Negro women will manifestly require a supervisor skilled in matters additional to archaeology. For that reason we have recommended the employment of Miss Christine Addock." Addock had previously participated in field and laboratory activities at the Museum; she was deemed a competent archaeologist. She had also "directed a white and Negro WPA crew here at the Museum cataloguing and studying material."[69]

Women in New Deal archaeology were conspicuous by their absence. Very few university-educated females saw activity in these formative years. Florence Hawley, trained at the University of Chicago in anthropology, figured among the most prominent specialists. No women, however, participated in the excavation activities of New Deal relief archaeology, although a scant number found employment in laboratory situations. The reader will recall the survey that Carl Guthe conducted in the pre-Depression era. It identified a dearth of females in archaeology; a significant percentage, though, comprised the rolls of university-trained individuals with graduate degrees in anthropology. James Griffin documents four women who achieved the rank of Fellow at the initial meeting of the SAA in 1935: Florence M. Hawley, Anna H. Gayton, Frederica de Laguna, and Katherine Bartlett. Archaeology in the postwar period successfully recruited large numbers of women, but this was not the case in the 1930s. No single reason exists to explain this phenomenon. One offered by the professionals of the period alleged that the difficulties associated with field life deterred many females. A second explanation suggests that since only black women in the South worked on the digs and the heavy labor requirements were construed as suitable only for "black labor," archaeological fieldwork was not something that respectable white women should do, regardless of their economic standing. Finally, it has been suggested that women were more attracted to the contemporary nature of cultural anthropology, not archaeology. The most influential women who exerted an impact on archaeology in the New Deal filled positions in the Works Progress Administration bureaucracy; none of them were archaeologists.[70]

DeJarnette's supervisors represented the Universities of Alabama, Kentucky, Arizona, Nebraska, Chicago, Denver, and Harvard. All of them possessed extensive field and/or laboratory experience. Again, TVA and WPA activities facilitated the dissemination of knowledge by encouraging professional archaeologists with varied backgrounds to interface. The result was that they refined the most popular field and processing methods of the time.[71]

DeJarnette encountered political complications similar to those faced by Matthew Stirling in Florida during the Civil Works Administration. The sheer size of the operations also brought labor problems, which in turn created unique challenges. The dearth of available labor became more troublesome as the decade progressed. The impending war in Europe and the Far East created more favorable economic conditions, and competition for labor became more pronounced. The belief of the local relief personnel that work of a more practical nature should take precedence exacerbated the situation.[72] Webb ultimately aided DeJarnette when he assigned directly to him a small labor force of Tennessee Valley personnel. A sympathetic posture here from the TVA permitted the supervisors to overcome a series of logistical problems. In fact, the Authority underwrote a substantial portion of the costs of equipment and salaries in the Alabama expeditions.[73]

Beginning in 1938 the Alabama relief program came under the scrutiny of the Washington Works Progress Administration. The issue: were proper scientific standards practiced? The archaeological consultants, Vincenzo Petrullo and, later, Stella Deignan, personally inspected the Alabama expeditions.[74] The National Park Service also sent over Arthur Kelly from Georgia. Kelly thought the projects of suitable quality, although he expressed suspicion about Walter Jones.[75] Alarm that the Alabama group might operate in Georgia and thus infringe upon his domain actually lay at the root of Kelly's reservations about Jones. Of legitimate concern, though, was the matter of laboratory processing of recovered material. The virtual inundation of the lab facilities by artifacts and skeletons, not an absence of planning, drew the attention of WPA officials. Negotiations resulted in the establishment of a laboratory in Birmingham staffed by sixty technicians. That operation became proficient in its duties and alleviated the worry.[76] The WPA also stressed the establishment and successful workings of the Laboratory in several press releases.[77]

A companion issue manifested in these digs stemmed from questions about standardizing the training of men who had no practical experience in archaeology. The staff created a manual that outlined detailed instructions on the duties expected of those both in the field and the laboratory. A later analysis of that document by James Griffin confirmed that it represented the essence of the Chicago approach.[78]

As World War II approached and funding slowly became more restricted, the Works Progress Administration, Deignan, and the Smithsonian placed more pressure on the Alabama crews to finish the work in the Pickwick Basin. Deignan maintained that no federal funding for additional work would be forthcoming until this was accomplished. She also awaited a final analysis of the laboratory data. On Alabama's behalf, the sheer numbers of artifacts produced were staggering. Processing time therefore lagged behind the pace established in other states, despite the contributions of the Birmingham laboratory. Ultimately, the directors accomplished all of these tasks. Any future conflict between state and federal sources ended when the United States declared war on Japan.[79]

Exactly when the expeditions might be completed likewise concerned both the Smithsonian and the National Park Service. The Tennessee River Valley faced immediate flooding just when federal monies became increasingly scarce. Both agencies worried that the Basins might be looted if the projects ended too far ahead of the influx of water. Hardly a new sentiment! Archaeologists had expressed identical anxieties for a number of years.[80]

Workers obtained information from the Alabama sites under unique circumstances, often at extra cost and effort. Some sites could only be approached by boat. The huge Pickwick Dam, some 7,715 feet in length and 113 feet high, was located in Hardin County, Tennessee. It created a lake about 75 miles square.[81] After a preliminary survey, the supervisors directed the excavation of 332 sites, with the Copper Galena of Copena Culture representing the most significant discovery. These island inhabitants on the river had been

Mississippian and had utilized the islands as a natural defense. Other data convinced Webb and DeJarnette that a pre-Copena focus existed.[82] Further information derived from the Wright mounds and village site, and the Colbert Creek site, presented substantive evidence for this idea.[83] DeJarnette defined other shell mound material from Pickwick as Archaic.[84] The shell mounds, a common type found by all Tennessee Valley Authority archaeologists, comprised apparently a static culture, with little intrusion of other groups.[85]

The other major salvage area investigated was located near Guntersville Dam and comprised twenty-three designated sites. Here, the archaeologists competed for labor with the Authority, as well as with local projects. Supervisors, not surprisingly, witnessed a marked decline in the numbers of available workers. Simultaneously, the amount of time to complete preservation attempts decreased in proportion to the rapidity with which the water level rose. Toward the end of the projects, crews had to excavate with bulldozers and grab what artifacts they could. Under those conditions, the field teams could not justly be accused of utilizing poor standards. The impossibility of traditional excavation meant that little beyond accumulation of artifacts could occur.[86] William Webb later classified the information gained from the Alabama Authority activities into five categories: an older pre-pottery period (Whitesburg Bridge); fiber-tempered pottery (from the outside); local limestone-tempered pottery; a group that buried its dead in the previous group's middens and which had shell-tempered pottery and truncated pyramids; and a final group, which was marked by European evidence.[87]

Those areas not covered by water constituted the remainder of the Alabama archaeologic conservation activities. An interesting site at Whitesburg relied totally on black men and women for labor.[88] One survey focused on Clarke County, along with excavations of mounds in Baldwin, Mobile, Madison, and others.[89] The Civilian Conservation Corps assisted with some of these efforts.[90]

These expeditions, of a true salvage nature, helped to create a larger conceptual view of the Southeast and the Upper South. A final legacy took the form of a strong archaeological presence in the University of Alabama with the appointment of David DeJarnette.

If Louisiana represented one of the more stellar examples of high-quality Works Progress Administration archaeology, then Georgia deserves the title of most publicized. The state received a shower of constant attention, with special focus directed at Ocmulgee and activities at Savannah. Press information for these areas far exceeded that given to the other sites. Even the activities of the consultant Vincenzo Petrullo were described in print. The WPA was consistent throughout its lifetime in bringing its story to the public.[91]

A greater significance resides beneath the veneer of media coverage, however. The Georgia experience played a prominent role in the further integration of archaeology's institutional elements. It also contributed to the growth of increased political sensitivities. Further, endeavors there mirrored the changes afloat in government archaeology. This likewise prompted action at the federal level by academics. In good part, all of this stemmed from the presence of the Smithsonian and the National Park Service. Added to this mixture was the interesting character of Arthur Kelly.

A paradox is inherent to this chronicle. The Society for Georgia Archaeology had worked diligently to bring quality archaeology to the state. Yet, within three years of the initial start-up of relief operations, control had slipped from their hands; a trio of non-Georgia groups now directed the destiny of the state's preservation crusade. In the other southern states, archaeology programs had only to address the Works Administration and the Tennessee Valley Authority. Conversely, for the Works Progress agency outside Georgia, most of its administrative headaches involved supplies of labor. But in Georgia, the WPA had to combat the entire state bureaucracy. The entrenched white power structure resisted the intrusion of federal relief into state politics at every level. In spite of the altruistic attempts at relief, federal programs were hindered in myriad ways, deemed as they were an incursion on the body politic. Fortunately, archaeology was only occasionally hindered by this circumstance.[92]

Recall that the initial Civil Works Administration actions at Macon did not rely upon a strong state framework for conducting archaeology. The Society for Georgia Archaeology became the first proximate statewide agency to coordinate work. Of course, it had just formally incorporated in the early years of the Depression. The federal activities here thus helped to accelerate the diffusion of a statewide academic apparatus, first through the National Park Service and later through the University of Georgia.

The Smithsonian, we know, had an officially designated role within the Works Progress Administration. That did not proscribe it from fostering a close relationship with the Society nor from maintaining interagency ties. It followed that the SI helped to guide the course of archaeology in Georgia. Frank Setzler played a central role. He aimed primarily to develop a statewide survey. Setzler likewise cooperated with Arthur Kelly; the latter aspired to extend his influence and status there.

Georgia, first through Kelly and then through Robert Wauchope, witnessed problems of personality similar to those Lewis brought to Tennessee. Kelly, too, expressed similar sentiments about the centrality of his stature. For one, he resented the attention given to other young archaeologists. He did not command the same respect as did Ford. Shadowed by others, he reacted to any change as a potential threat to his reputation and authority. The fact that his skills as an archaeologist were not especially strong reinforced such behavior. His extensive correspondence with friend and ally Setzler described a series of petty difficulties. He sought constant reassurance from Setzler, who answered the lengthy letters patiently.[93] His insecurity often manifested itself in the inability to effect affable relations with his subordinates. Still, the Georgia digs were placed under his central guiding hand and accomplished significant work. Kelly's role within the Park Service, combined with close Smithsonian ties, meant that archaeology in Georgia was rooted in a broad federal base. A strong federal presence for affecting decision-making was equally obvious.

Supervisors arrived and almost as quickly departed, for few could collaborate with Kelly. Vladimir Fewkes, an established archaeologist with stronger academic credentials, came down to work in Georgia for a time. Before the two even met, Kelly forwarded a letter telling Fewkes exactly how the digs would be organized and run. The terse tone of the letter was not conducive to the creation of a smooth running, professional relationship.[94] In another case, Joseph Caldwell, who performed so well in other WPA expeditions, Kelly considered "too immature."[95]

Regardless of Kelly's shortcomings, the inescapable fact remained that the story of Georgia archaeology revolved around him. Consequently, the Park Service acted as the essential institutional link in the exercise of relief archaeology, a circumstance further solidified when the Macon works became a national monument. Thus, the NPS was the linchpin in a system that saw archaeology in this area of the South incorporated into a larger national structure.

In spite of the appreciable problems evident in the peculiar Georgia setting, matters of labor did not generally plague Kelly. His favorable contacts, both in the society and the federally administered state relief mechanism, precluded this. Later, his powers expanded beyond those related to the directorship at Ocmulgee with an appointment as superintendent of a new Civilian Conservation Corps camp.[96] On one administrative level he now controlled archaeological endeavors, on another the development of surrounding park facilities. Eventually, Kelly directed more than seven hundred workers, ranging from archaeologists and engineers to shovel men. With so many people to supervise, the situation on occasion became unmanageable. However, according to Gordon Willey, himself a participant there, relief laborers preferred this form of duty. Otherwise, their other options ended in road building or swamp draining. Most were understandably cooperative.[97]

If the persistent headaches about labor experienced elsewhere failed to materialize in Georgia, complications inherent to laboratory processing did. Ocmulgee and other Georgia digs produced a huge stockpile of artifacts. As a result, analysis and publication lagged behind data collection. The reasons were neither unique nor unfamiliar. Lack of trained personnel played a role. Laboratory work required appreciable background and skill. The other part of the problem lay with the Works Progress Administration. Criticism of the agency stemmed directly from its not providing enough money and satisfactory quantities of labor for this task.[98] There were several reasons for this deficiency. The archaeological projects comprised but a small part of a statewide program for relief. Each state, therefore, had to decide about the optimum manner for allotting limited funds. Certain types of projects inevitably received higher priority. In addition, state agencies assigned urban-based projects preference over archaeological laboratory work. Considering the geographical (rural) and intellectual scope of archaeology, it fared quite well when viewed in the overall scheme of things. The complaints raised by archaeological supervisors simply echoed those aired by directors of other forms of relief. Only so much money was available.

Kelly also expended considerable energy to overcome the strong local feelings about outside intrusion. This constituted another paradox, for Kelly simultaneously resisted what he perceived as interference by other agencies. He taught night classes for the workers and spoke regularly in front of a variety of public and fraternal groups. In doing so, he carried the message of archaeology to a large sector of the public. His generally competent management of the Ocmulgee Monument additionally served to create in the public's mind a conscious relationship between professional expertise and good archaeology. Press coverage was commensurate with the size of the Macon operation. Newspaper accounts of the daily total of artifacts gathered and the numbers of persons involved created long-lasting images. Articles likewise recounted the visits of notable researchers such as Vladimir Fewkes.[99]

Four separate projects occurred in Georgia: Ocmulgee, Chatham and Glynn Counties, and the state survey. Workers collected and stored a huge amount of material, some of which today rests unanalyzed. Assistant directors and supervisors authored the published accounts of the amassed data.[100] Kelly only penned a preliminary report and never completed that last responsibility.[101] Overall, the expeditions produced a considerable number of flints from a pre-pottery stage, followed by a pottery and agricultural element, and then the mound culture.[102] Preston Holder worked on the sites in Glynn County before departing for Columbia University to undertake graduate work. Work also focused on St. Simons Island, just prior to the construction of an airport there in 1937.[103] Holder also participated in the effort at Irene Mound with Vladimir Fewkes in February 1938.[104] This mound in Chatham County near Savannah became the shining light in the Works Progress Administration, paralleling the prominence relegated to the Macon Mounds during the Civil Works Administration.[105] Holder and Willey also worked solely with black labor forces, in one case, all black women. Georgia utilized black labor for reasons described previously. Excess labor pools of unemployed blacks existed. Political pressures emanated from Washington to utilize blacks when possible. However, an additional factor related to the supervisors, according to Gordon Willey. Northern and western archaeologists had no qualms about working with blacks. This last point is again illustrative of the impact exerted by the individual site directors and their supervisory staff.[106] These projects, albeit examples, demonstrated that direct opportunities did exist for American blacks in the New Deal. Fewkes, in one instance, both expressed gratification for the support he received from the community and commented favorably on the employees. He further observed, "We have 117 people in all, among them eighty-five colored, of whom seventy-nine are women."[107] The scene was not remarkable in that blacks were employed, but because they toiled alongside whites. Segregation stood as the usual procedure for the southern relief projects, but this occurred too around the country. The tone and manner of supervisors' perceptions of black workers, however, were not generally the same as that applied to whites. Recall how W. Duncan Strong had praised his crew in California during the CWA. In 1938, Lucy McIntyre, a WPA employee, discussed black workers before the Society for Georgia Archaeology in Athens. The title of the address was "The Use of Negro Women in the WPA at Irene Mound, Savannah." The project utilized about 100 black women. The WPA felt this archaeological project was an ideal situation in which to utilize unskilled labor. The workers were "careful, docile, efficient workers especially suited to this type of work." Their "childlike interest" kept them on the job, "singing as they worked." Of course, "an intelligent class of white foremen supervised." In Strong's opinion, his California workers were motivated by the thrill of discovery. Here they were childlike.[108]

Robert Wauchope supervised the survey, which when completed provided a comprehensive view of the status of Georgia archaeology for the first time.[109] Wauchope wrote generous words for his workers but expressed disdain for the bureaucracy. He had been appointed to conduct the survey after a lengthy interview process. For whatever personal reasons, Wauchope expressed annoyance with the peculiarities of the administrative process. He groaned that he had to account for every dollar and man-hour. It constituted an unnecessary drain on his time, he added. His monthly reports exuded an almost venomous feeling about this subject. Of all the reports on deposit in the National Archives or at the National Anthropological Archives, his stand out as the most negative. The remainder of the archaeological community seemed to understand the concept: in a time of depression, for the money and the opportunity, one had to account for it. The price appeared to be a small one. Bureaucracy could claim legitimate needs. Such a widespread program carried the possibility of corruption at every turn. Georgia state administrators had already demonstrated that all too well. Paperwork was simply a part of that overview process. In Wauchope's case, however, the intensity of those vitriolic expressions seems largely unexplainable. The dividends of a state survey far outweighed the admitted inconveniences. Consider also that academic archaeology achieved permanent establishment in Georgia as a direct result of the presence of federal relief work. Later, this very venue spurred employment for both Kelly and Wauchope at the University of Georgia.[110]

Unquestionably, academic archaeology benefited from interaction with federal agencies. Significantly, most archaeologists associated proper or scientific archaeology with the activity of university-trained professionals. Since these individuals then resided almost exclusively in colleges, museums, and government, the status of the archaeology-university connection grew considerably. The high degree of success associated with the relief investigations reinforced the notion of college-based archaeology as the preferred mode of inquiry. Any future work, either in state or federally directed archaeological efforts, relief or salvage in nature, would include "professional" archaeologists.

In a twofold manner the positive reception by the public assured archaeology's station outside the government realm. First, on a continual basis for over seven years, Works Progress Administration publicity campaigns effectively communicated information about professional caliber archaeology. In those areas that already exhibited a high degree of public interest, archaeology enjoyed yet a more elevated status. Second, the expenditure of local funds, geared either to supplement the research process or simply to bolster the tourist trade, reinforced notions about the validity of the "new" archaeology.

Another important byproduct of the 1930's experience was the expansion of the size of the professional community and its communication system. New departments of anthropology merged into the nascent national network. Participation in many regional programs grew, the now popular conference system at the forefront.

Relief archaeology aided the discipline within the university structure as well. Availability of federal monies spurred the establishment of departments and chairs in anthropology and archaeology. Individuals obtained requisite funds for research and opportunities for publication. As the need for more trained field personnel expanded, student enrollment likewise increased. Relief project endeavors also supplemented traditional teaching processes since many graduate students worked as supervisors on the expeditions.

Conclusion

By 1939 archaeology's leadership acknowledged the many accomplishments of New Deal archaeology. They also understood that significant problems remained. The lack of consistent quality production under the Works Progress Administration and the ongoing reduction of federal expenditures stood foremost. Regarding the latter, the community faced a financial crisis in some ways similar to that experienced with the inception of the Great Depression. An uncertain future included the federal government planning a host of new postwar projects that would inundate especially sensitive archaeological environments. Without doubt, this would also necessitate greater communication within the field at state levels. Additionally, the public, the Congress, and other powerful federal agencies required education about the scientific and historical importance of the nation's archaeological heritage. The potential to act in a concerted fashion seemed very likely, given the accomplishments of the previous twenty-five years. Academic archaeologists had staffed more of the government bureaucracy. The size of the professional community throughout the country had increased in colleges and universities, museums, and societies after 1937. A discipline with many components required only collective action and decisive leadership.[1]

Between 1939 and 1941 these professionals acted to establish a political and institutional framework that would determine the content and course of their research. As before, agencies within the federal government, specifically the National Park Service, the Smithsonian Institution, and, for a short time, the Works Progress Administration, remained the critical entities within this realm.

Carl Guthe aptly described in 1939 the concerns of many archaeologists. His *Science* article focused on the basic needs of the discipline.[2] It also related how the experiences and insights gathered during the 1930's relief experiments dominated their thoughts. Through their work, archaeologists learned the value of utilizing government funds for exploration. The huge amount of recovered data made comprehensive interpretations feasible. Pivotal leaders like Guthe viewed the discipline's problems on a national scale. Considerable concern emerged that they might be excluded from any new decision-making processes. A firm belief in the soundness of their approach to the study of archaeology was almost universally accepted. A strong faith pervaded the professional community respecting future opportunities. An increased sense of unity in the academic community arose from the collaboration made necessary by the relief programs. The limited size of the discipline made select professionals more visible and influential. A zealousness for continuing reforms begun in the 1933–38 period abounded. Finally, the archaeological leadership lacked full confidence that government agencies could provide the essential leadership in decision-making.[3]

Briefly, the initial attempt to redefine policy-making again began within the WPA. Florence Kerr, head of the Women's and Professional Division, moved to include professional archaeologists in the decision-making network. She contacted Ross Harrison at the National Research Council. He, in turn, gave Kerr's letter to Carl Guthe. The former director of the State Surveys Committee formed a committee to deal with the specific problems inherent to relief archaeology. The group was known as the Committee for Basic Needs (CBN) on American Archaeology. W. Duncan Strong (Chair), William C. McKern, William S. Webb, J. O. Brew, Clark Wissler, Alfred V. Kidder, and Fay-Cooper Cole comprised its membership.[4]

The committee was affiliated officially with the National Research Council through its Division of Anthropology and Psychology. None of the members represented any federal agencies. It was hoped that this stance would limit governmental influence.

Convened in May and June 1939, the initial meetings of the CBN produced fruitful results. Based upon discussions of Works Progress Administration relief work, a consensus was reached as to what in the future would be defined as scientific archaeology. The committee disseminated the confidential reports of these meetings to the larger community through an article in *Science* magazine, not *American Antiquity*.[5]

The article addressed several aspects of Works Progress archaeology. Its recommendations specified guidelines for the definition of archaeological problems, standards of field and academic training for personnel, standardized procedures in laboratory analysis, and timely publication of data. Criteria relative to training, field procedures, and publication had been present in archaeological procedures for some time. The definition of what constituted an archaeological problem represented another matter. Archaeologists henceforth would have to think in terms of huge cultural areas. The article concluded that failure to implement all suggestions would diminish the scientific worth of future research. The committee thus urged that all levels of government refrain from subsidizing any work not meeting these conditions.[6]

The last two formal meetings of the Committee on Basic Needs took place in December 1940 and March 1941. The minutes of the convocations indicate that Stella Deignan wished to incorporate the CBN into the WPA approval process. The committee had achieved its ultimate goal: to influence decisions about archaeology in government but without undue federal bureaucratic interference. Deignan even suggested that the committee's members evaluate the suitability of future relief personnel, thus creating a file of qualified directors and supervisors.[7] In reality, this would have formalized the casual referral system already in place, only more universally. Considering the breadth of the operations, the file would be large. Importantly, it would only list those individuals qualified to carry out scientific archaeology. Such a policy would relieve Deignan and her agency of some of the responsibility for which she and others lacked proper training. For example, she had a degree in anatomy, not archaeology. Her role progressively became that of a political liaison rather than a professional consultant, although some of the duties of each role overlapped. During the remaining months of the agency, the CBN and Deignan devoted most of their attention to the subject of publication.[8]

Absence of significant funds and U.S. entry into World War II ultimately precluded implementation of this strategy. WPA headquarters issued an official termination notice relative to remaining relief programs in January 1942.[9]

By mid-1941, the WPA and other related relief agencies had had their programs drastically scaled back. The U.S. economy was growing in direct response to increased spending relative to war sales and production. Still, a groundwork for future action had been laid. In July 1944 the need to respond stimulated the creation of another group, the Planning Committee. Their expressed function was to

> consider the status and disposition and use of the archaeological data, reports and objects secured by the various archaeological projects of the late WPA. Later they will consider, in connection with the Basic Needs

Committee of the National Research Council, any other policies pertinent to the welfare of American Archaeology, or of this Society.[10]

After considerable consultation and research, an index of the reports was created. The reports themselves were placed on permanent deposit in the Smithsonian's U.S. National Museum.[11]

In 1945 the Planning Committee learned of an announcement by the Army Corps of Engineers for postwar plans of dam construction. The eventual discussions and political maneuvering would take the form of a cooperative agreement between government agencies and professional archaeologists.[12]

The eventual demise of the Works Progress Administration made fulfillment of the Planning Committee's mandate all the more imperative. The relief agency, irrespective of the problems that plagued it, had at least functioned as a single, unified entity capable of affecting collaboration with the profession. However, a major source for decision-making and funding disappeared when Congress dismantled the WPA. Efforts by professional leadership eventually helped to create the Committee for the Recovery of Archaeological Remains (CRAR) in May 1945. It worked with various political and administrative entities in Washington to create a position for archaeology in planned river-valley construction programs. The Smithsonian joined with the Park Service, the Bureau of Reclamation, and the Corps of Engineers to carry out extensive salvage operations.[13]

The committee's purpose was to create an independent body devoid of reliance on any relief agency. In their words, archaeology would "stand on it own feet, the only objective being the scientific one. It is not encumbered by relief problems or any other unrelated condition."[14] The CRAR would become the prominent political means for archaeology to interact with federal agencies in the next generation.

An interesting footnote to this discussion merits comment. In 1939 the activities of a group led by James Griffin indicated that the members of the discipline had already expressed thoughts similar to those that culminated in the organization of the Committee for the Recovery of Archaeological Remains. Griffin's group envisioned the development of multifaceted cooperative endeavors to elevate the status of scientific archaeology.[15] Griffin, Phillip Phillips of Harvard's Peabody Museum, and James A. Ford of Louisiana State University submitted a proposal for an archaeological survey of the Central Mississippi Valley.[16] They observed that recent archaeological work had uncovered tremendous stores of information relative to almost the entire region east of the Mississippi River, as well as a considerable amount on the Plains from Texas to the Dakotas. Extensive research also extended to the Upper and Lower Mississippi. They surmised that, elsewhere, thorough comprehension of an entire prehistoric cultural framework had not been attempted. In a well-prepared document, these three aimed to survey the area.[17] Instead of channeling the proposal through the Smithsonian, they requested aid and assistance from the National Park Service and the Civilian Conservation Corps. A subsequently revised draft made clearer their intention to enlist the cooperation of Harvard, Louisiana State, and the University of Michigan. By this time, the group also included Gordon Willey from Columbia.[18] According to Griffin, it seemed preferable to coordinate the wide-scale survey through the National Park Service. They thought the Works Progress Administration an inappropriate agency to assist in the task because of its state-structured orientation. The Smithsonian, due to the considerable strictures regarding budgeting and personnel that governed its operation, likewise seemed an unattractive alternative.[19]

In late 1939 and early 1940, then, elite members of the archaeological community perceived that the community's larger needs could no longer be met by either of the two extant government organizations who had previously provided that service. Later actions by archaeology's leaders seemed not so incongruous. Interestingly, they opted to rely on *another* federal agency, here the NPS. American archaeology's long liaison with the federal government, though strongly supplemented by the rising preeminence of its university and college component through the 1920s and 1930s, still found its destiny tied to its original benefactor.

Epilogue

The story of American archaeology in the New Deal gives another dimension to the Depression chronicle. Too often, its history has been reduced to an exposition of the amount of dollars spent or the number of people employed. Labor certainly benefited from the Wagner Act, retired people from Social Security. In the case of archaeology, the relief programs exerted a less immediately visible or discernible impact.

In 1948 Walter W. Taylor wrote for his doctoral dissertation at Harvard University the first monographic critique of concept and interpretation in American archaeology.[1] To Taylor, historical sequencing of cultures by necessity preceded other forms of archaeological analysis. He next specified what he construed as the proper outline for conducting and interpreting archaeological research.[2] Taylor then proceeded to criticize the current state of American archaeology and those responsible for its development. Chiding them for placing too much emphasis on classification and taxonomic study, Taylor attacked most of the major archaeologists discussed in this monograph.[3] He lashed out at A. V. Kidder for identifying his research as theory construction when it resembled chronicle.[4] Emil Haury, Frank H. H. Roberts, William Webb, and J. B. Griffin all placed excessive emphasis on classification.[5] Utilizing blunt critique as his vehicle and advocating carefully applied analysis of the data, Taylor summarized his conjunctive approach. In this scheme, the archaeologist viewed all possible paths of inquiry as a unified entity in order to understand problems in their totality. Data would be drawn from every extant source, ranging from artifacts to statistics.[6]

Taylor perceived the signals of change then afoot in archaeology. However, the reasons for the phenomenon eluded him. Emil Haury answered this charge when he wrote in 1985:

> The penchant to classify and gather facts during the second quarter of this century ... represented a significant step in the maturing process of the discipline and created the platform from which further advances could be made.[7]

Taylor lacked a definite historical sense of the transformation that had occurred in his discipline. His synthesis was devoid of the very understanding he accused the other archaeologists of lacking: conjunctive functionalism. Archaeology found itself on the threshold of intellectual and institutional change; this circumstance evolved from a variety of interactive forces. Taylor recognized the essential step of gathering data in order to create larger conceptions, either classificatory or taxonomic. Upon reaching that point, however, the archaeologist had to use anthropological theoretical tools to analyze the data for the information it contained about the dynamics of culture.

Taylor had created a model that closely resembled anthropology. But he failed to comprehend that archaeology had not followed a similar cognitive evolution as had the parent discipline. Taylor appears correct in having called for interpretive work. After all, archaeology had finally attained a position, in terms of its institutional development and available data base, to offer higher level regional interpretations. Taylor failed to grasp that the interpretive role now fell to him. He belonged to the generation in the best position to do just that. He could stand on the shoulders of those he criticized, the professionals who had transformed the discipline from the strictures of nineteenth-century Baconian science. Taylor should have compared their work to that carried out in the 1800s in order to ascertain the evolutionary nature of archaeologic investigation. The classifiers completed their task only after considerable effort. Their achievement perhaps stands as the greatest legacy of Depression-era archaeology, that is, the fertile institutional, professional, and interpretive opportunities it bequeathed. Of the notable achievements of the postwar period were the many attempts at interpretation. Taylor had prophesied the future but misunderstood its antecedents. Broad interpretations surfaced, not because the past leaders had failed to create them, but because they did not have the conjunctive advantage of the present.

Perhaps no better indicator exists that attests to the efficacy of the viewpoint presented here than a comment offered about a seminal book in the history of American archaeology. Carl Guthe, in his 1967 article, "Reflections On The Founding Of The Society For American Archaeology," observed:

Gordon Willey's concomitant success in combining the archaeological evidence gathered from all parts of the Americas into a single and coherent statement of the prehistory of the Western Hemisphere is a most astonishing achievement. Such a synthesis could not have been contemplated when the Society was founded.[8]

Guthe, as usual, demonstrated insightful perception. The latest generation of professionally trained academic archaeologists had conceived not just of a single region or state. Instead, they had envisioned the archaeology of a continent and beyond. Willey and Jeremy Sabloff elsewhere had also chronicled and attempted interpretation of the methodological changes evident in American archaeology; in other words, they had attempted to provide a narration of the development of archaeological scientific procedure.

William G. Haag six years earlier had made a similar reflection while attempting to identify the pivotal turning point in the development of eastern archaeology. He cited a 1948 comment by Guthe, identifying the appearance of the Midwestern Taxonomic System as the turning point.[9] Haag disagreed. He thought the introduction of the Carbon-14 dating method was the vital factor. From it, archaeologists could finally detail accurate temporal classifications.[10] Both were likely correct. The Taxonomic Method constituted a leap forward for eastern archaeology. It helped to stimulate discussion aimed at construction of larger, more accurate, classification systems. The radiocarbon dating method had rapid application around the country. But archaeology had already witnessed classification efforts in the Southwest with the application of dendrochronology and strata-dating procedures.

In an institutional sense, archaeologists reached a watershed when they perceived themselves as somewhat distinct from their anthropological brethren. Consequently, they sought different organizational strategies to serve their needs and to protect and expand the power of academic, professional archaeology. The founding of the Society for American Archaeology represented one symbol of that process. It had begun earlier, with the State Surveys Committee, and it came to fruition with the formation of the Committee for the Recovery of Archaeological Remains. The New Deal made that possible, and more, by fostering the development of huge new masses of data available for future researchers, particularly in the eastern united States. The notion of a scientifically oriented, professional archaeology was transmitted through relief laborers, appeared in government press coverage, and was symbolized in new museums. Tourism only further enhanced the ongoing historic preservation movement. Both recently organized and long-established museums received much of the material. Staffs of university-trained professionals arrived soon thereafter. The broad academic community established greater communication networks with the reaffirmation of the worth of regional conferences, which also increased in number. Expanded publication opportunities for more societies and government agencies complemented this movement. The growing demand for departments to include training in archaeology in colleges and universities resulted in still more job openings. Similar demands for recent graduates surfaced in government agencies, as preservation responsibilities diffused throughout the states. Extended funding opportunities resulted from supportive government legislation. Greater numbers of archaeologists filled foundation positions and scholarship boards. The political actions of professionals supplemented those of the government agencies. Strategies were developed to influence the legislative process and helped to insure a continuing favorable climate for this archaeology.

Archaeology formalized its approach to training and analysis. Larger interpretive efforts resulted. Training had become more standardized, as had language, terminology, and taxonomic systems. A greater pool of trained people with greater facilities, synthesized knowledge, and new data made larger temporal and interpretive sequences possible. This facilitated a wider frame of understanding about the North American Indian cultures, not only in a focused sense but in a more expansive regional or multiregional purview as well. Ecological and environmental knowledge increased with archaeology's continued reliance on other specialists; expeditions became even more widely collaborative. The experiences of the 1930's generation of archaeologists assured a continued emphasis on fieldwork, but accompanied by the expansion of laboratory analysis and publication. Lastly, archaeology continued to depend on government, with its inherent political turmoil, agency rivalry, and fluctuating budgetary process. The profession consequently had to refine continuously its level of political sophistication, including the need to improve its public image.

Appendix
I
FERA: Private and Other Measures

The projects of the Federal Emergency Relief Administration continued the policy followed by the Civil Works Administration: funded excavations and surveys would last only a short duration. Still, these projects held a special significance, functioning as a training ground for graduate students throughout the country. Utilized in supervisory, laboratory, and field capacities, individuals gained first-hand experience at a variety of sites, even if they were not of a large-scale, cooperative nature. Sometimes, participants received pay, at other times they simply fulfilled academic requirements. These young men often published material drawn from that work. Many later returned to the university to pursue a doctoral degree. Gordon Willey and Emil Haury, for example, followed such a path to high academic standing. Several of these persons moved into positions of leadership in the archaeological community after World War II. Additionally, work continued under the auspices of several other groups, private and governmental, all of which furthered the growth of the data base for archaeology.

Montgomery County, Kentucky, hosted the largest state Relief Administration site. Dr. W. D. Funkhouser conducted the project for the Department of Anthropology and Archaeology. He directed the excavation of a mound that contained artifacts and skeletal remains similar to Algonquin or Woodland culture. Artifacts of copper, stone gorgets, and clay indicated a potential northern influence.[1] The other southern expedition at Lake Pontchartrain, carried out by Louisiana State University, had begun prior to the birth of this first New Deal agency. An extensive collection of potsherds disclosed some forty-eight different decorative patterns. Both sites in this dig revealed important information that linked this area with Woodland groups in other parts of the state.[2]

The excavations at Macon continued under the direction of A. R. Kelly. Results were included in a later report, in which Kelly compiled the information gathered from 1935 to 1938.[3] The Society for Georgia Archaeology remained active and worked for the creation of the Ocmulgee National Monument. The members also led the movement to establish a chair of archaeology at the University of Georgia.[4]

Smaller expeditions included those in Mississippi, Missouri, New Mexico, and Oklahoma. These digs, and surveys around the country, were important. In spite of their limited size, they sustained the growing liaison between the government and archaeology. Equally meaningful, the notion of relief archaeology spurred the dissemination of an increasingly complex scientific archaeologic methodology beyond the South.

In Mississippi, the State Archaeological Survey conducted fieldwork sponsored in part by its Department of Archives and History and the Smithsonian. Moreau B. Chambers, aided by two student archaeologists and Relief Administration labor, excavated two sites and collected specimens from others in Clay and Okitibbeha Counties, an area lying between known Chickasaw and Choctaw lands. The digging revealed burials, sherds, kitchen middens, and artifacts similar to surrounding cultures.[5]

The University of Missouri instituted an archaeological survey of Missouri, with Professors Jesse Wrench and Brewton Berry as directors. The survey accounted for a minimum of 13,245 mounds, 950 village or camp sites, and 168 caves. In addition, a census of persons interested in archaeology and those with collections was assembled in order to acquaint the investigators with the material already extant.[6]

A small expedition in Beaver County, Oklahoma, disclosed a number of empty graves and some remains of human skeletons, along with a few artifacts. The early inhabitants apparently migrated to this site from Kansas. They then moved further on into New Mexico. The evidence gathered facilitated understanding of migration patterns in this area.[7]

The University of New Mexico and the Museum of New Mexico carried out five expeditions with Emergency Relief Administration aid. The work, done in a great Kiva–type and a small house (Talus Unit No. 1) in the Chaco Canyon area (three at Chetro Ketl), included a spectacular find of nine complete ceramic vessels. This program also functioned as a field school for ten graduate students.[8] Another annual summer affair, the Jemez Field School, participated with an enrollment of fifty students. An additional project near Santa Fe, New Mexico, contributed new information and material amenable to tree-ring dating analysis.[9]

It is important to remember in any discussion of New Deal archaeology that non-federal agencies instigated work. Carl Guthe's State Surveys Committee recorded forty-one federal and non-federal activities current in 1934 and 1935. M. R. Harrington of the Southwest Museum explored the Pueblo Grande de Nevada near Overton, Nevada. The State Park Division of the National Park Service conducted the work and shared the collections with the Museum. Two companies of the Civilian Conservation Corps, the 974th and the 573rd, supplied the labor.[10]

Oregon witnessed activity similar to that in the Tennessee River Valley. Herbert W. Krieger of the U.S. National Museum worked the Columbia River Gorge. The construction of the Bonneville Dam was destined to flood a portion of the river basin. A Public Works Administration allotment paid for a share of the work.[11] The National Park Service sponsored a small operation in Utah in the Zion National Park, while the U.S. Indian Service and Brigham Young University carried investigations into the Utah Basin. Albert B. Reagan began the endeavor with the Indian Service but completed it as a special professor of Anthropology at Brigham Young.[12]

Considerable activity in Texas centered around Dr. Stuart Johnston, of the newly established Department of Archaeology at the West Texas State Teachers College. This initiated a period of expeditions that extended into the next relief era.[13] Several private foundations, most notably the Carnegie Institution of Washington, continued to sponsor this effort.[14] Private money was of significant importance in the early 1930s as governmental funds diminished.

Several of the programs during this period conducted surveys of the various cultural regions or states. Established earlier, this approach occurred prior to initiating larger scale efforts. Subsequently, these surveys quite often found final form in the Works Progress Administration. The Nebraska State Historical Society continued survey work with Waldo R. Wedel of the University of California, assisted by five volunteer students. This effort augmented the work of a talented Historical Society amateur, A. T. Hill. Likewise, the University of Nebraska survey endured with Earl Bell. This institution continued a tradition of archaeology and anthropology originated by W. Duncan Strong before he moved on to the Smithsonian.[15]

Cooperative arrangements between expeditions also existed in this period. For example, the privately financed archaeological work at Gila Puebla and the Laboratory Training Course in Archaeological Field Method associated with the Laboratory of Anthropology arranged a joint agreement. Emil Haury, the Assistant Director at Gila Puebla, carried cooperation to fruition with a course he offered in New Mexico for six scholarship recipients.[16] Private archaeology in Arizona and New Mexico provided considerable opportunities for professional archaeologists and students alike throughout that region.

Private archaeology not only functioned as an alternative means to retrieve archaeological information, it also operated as another avenue to achieve status. Conscientious amateurs, often talented in their own right, made important contributions either through self-financed efforts or efforts sponsored by government. The existence of both attested to the persistence of pluralistic, if also divisive elements in American archaeology. This hampered the campaign by the academic archaeologic segment to wrest control of the discipline. Yet, at a time of severe economic trouble, private archaeology constituted an alternative source of funding. Still, the notion lingered that it remained possible to pursue such knowledge in an avocational as well as a professional capacity. Academic archaeologists struggled to overcome the deleterious consequences of their association with amateurs. They continued active on this front from the 1920s and into the post–World War II era.

Interestingly, the bulk of the membership rolls in archaeological societies around the country was comprised of non-university-trained people. The discipline wished to monitor and control this element of American archaeology. But it was often obligated to rely heavily on amateurs for financial and logistical support. The exploits of the members of the Society for Georgia Archaeology, not to mention those of William S. Webb, have already been discussed. Harold S. and Winifred Gladwin worked as their counterparts in the Southwest.

A transplanted, successful eastern stockbroker, Gladwin exhibited myriad intellectual interests in subjects ranging from archaeology to butterflies. He shared them with Winifred McCurdy, later Mrs. Gladwin. She derived her wealth from inherited railroad money. A chance meeting with A. V. Kidder sparked an interest in the problems of Southwest archaeology. As a result, the Gladwins founded the Gila Pueblo expedition. Well established with facilities and a commitment to publication, this privately financed project provided substantial opportunities for many graduate students to gain field experience, including Emil Haury.[17]

Gladwin also contributed to the discipline in another manner, by offering his own culture classification for the Southwest, one as broad as the earlier Pecos Classification. Willey and Sabloff concluded that the system stood as one of the better-developed for that time. Gladwin's system, albeit constructed from a non-academic orientation, forced professionals to reconsider sequencing notions and aided in stimulating the refinement of regional interpretation.[18]

The Gladwin expeditions heralded significance in yet other ways. First, Gila Puebla enabled Emil Haury to continue his research. Other young students similarly gained advantages from the extensive training in fieldwork they received.[19] A well-sustained project of considerable duration, it also facilitated the disclosure of tremendous amounts of information. Moreover, it maintained an emphasis on fieldwork in yet another venue, that of the privately financed dig.

The experiences of the private expeditions revealed that the paths leading to the practice of archaeology still varied. The career of Emil Haury represented a case in point. A combined interest in natural science and geology, a family background in the humanities, and a driving curiosity were traits he brought to graduate school at the University of Arizona. Between 1927 and 1930, after receipt of his M.S. in anthropology, he furthered his education on Smithsonian expeditions. He also collaborated for a year with A. E. Douglass at the Flagstaff Museum. In those twelve well-spent months, Haury refined his laboratory skills, as he learned about tree-ring dating. He then accepted an offer from Gladwin to participate in the Gila Puebla venture. But to take that position, Haury had to reject offers from both the Smithsonian and the University of Arizona. During the seven-year-long apprenticeship, which encompassed the period 1930–1937, Haury also pursued his doctorate in anthropology at Harvard University. After conferral of his Ph.D., he returned to the University of Arizona to begin a prestigious career.[20]

Appendix 2

Further WPA Work: Institutional Transformation

This discussion of institutional development within a state, and on occasion incorporating a regional context, focuses on the further establishment of archaeology in the universities. Relief archaeology, as noted, worked at levels other than the national one. In turn, this precipitated subsequent changes in the disciplinary setting. Matters presented here include the actions of critical individuals, cooperation between local and government agencies as well as a variety of other institutions, new classification and temporal systems, accompanying economic benefits, and expanded public relations. New Deal relief programs exerted a pronounced impact on the structural growth of academic archaeology. The degree of impact depended on the stage of institutional development relative to each state; some schools had only recently initiated archaeological instruction while others were already moving toward inter- and intrastate cooperative endeavors. State projects, primarily channeled through established academic institutions and societies, reinforced the leadership status of the professional. The intention here is to complete our portrait of WPA archaeology, at times offering analysis of marginally tangential issues. The considerable variations in activity attested to the uneven development of institutionalized archaeology in many areas of the country. Nonetheless, the relief efforts facilitated integration of disciplinary elements within the fragmented community structure. Private and museum archaeology became more closely affiliated with its academic-based counterpart. Sometimes federal funds spurred this occurrence, other times the influence was indirect. It should also be noted that some archaeological research transpired without benefit of relief monies. Several states, those usually in the New England area, for example, had long conducted work through university and museum networks. Although hampered by the effects of the Depression, they sustained such efforts. While the institutional and structural changes in the discipline encouraged by the heightened federal presence lessened the impact of private archaeology, a strong presence remained in certain areas. In the Southwest, the private tradition ran strong.

A more advanced stage of the discipline's growth centered in the Arizona–New Mexico section of the Southwest. The events specific to archaeological investigation in the two adjoining states in some respects paralleled steps in the coalescence process then present in the eastern archaeological zone.

The cooperative effort in Arizona revolved about the University of Arizona. This institution, already active in Southwest archaeology, had by the 1920s offered a graduate program in anthropology with an emphasis on archaeology through the master's level. Expeditions that examined the many cultures there symbolized an ongoing theme of creating temporal sequences begun earlier by A. V. Kidder. In addition, the Museum of Northern Arizona, the Pueblo Grande Laboratory, and the large, privately funded archaeological expedition at Gila Pueblo thrived. A high degree of cooperation already existed between these agencies when the Works Progress Administration came to life. For example, the WPA monthly reports for Arizona officially listed the University of Arizona as the primary sponsor of state projects, with Roy Lasseter identified as the chief supervisor. However, a notation made in each report and the accompanying index specified that federal work in Arizona embodied a cooperative effort. The other agencies already mentioned also joined in the project. Additional supervisory personnel included Dr. Harold S. Colton, Dr. Emil W. Haury, Odd S. Halseth, and Harold S. Gladwin.[1]

Unique environmental conditions in the region allowed for the enhanced preservation of many types of artifacts, flora, and fauna, in particular, trees, given their importance for dendrochronology studies. The availability of tree-ring dating ushered in a more advanced, precise temporal classification than existed in the rest of the country. With a chronological system, easily accessible artifacts, and widespread cooperative movements, interpretive works appeared here earlier than in other parts of the country, the Northwest, for instance.[2]

Arizona's single WPA project commenced operation in October 1938 and lasted until June 1940. Funded with almost $100,000, this ambitious undertaking encompassed most of the state's geography. It included plans for appropriate mapping and preparation of charts and collections for later display. The basic work, conducted in Maricopa, Coconino, and Gila Counties, produced several hundreds of thousands of artifacts that served to illuminate the ongoing work initiated at Winona Village. Site work revealed at least four major groups: Mogollon, Hohokam, Pueblo, and Payayan. The huge supply of new information further stimulated cooperative work, as it spurred the convocation of more regional conferences. Meetings in 1931, then in 1936, built upon those of the 1920s. Frank H. H. Roberts remarked in a 1937 publication that:

> Never before in the history of southwestern archaeology has the field work of various institutions been spread over such a wide range of time phases as has been reported for the past year.[3]

Other forms of cooperative ventures, some governmental, marked the New Deal period in Arizona. In 1938 the Arizona Anthropological Association, in conjunction with the City of Phoenix and the City of Globe, supported work at the Pueblo Grande Laboratory. In 1939 the University of Arizona carried out some research on the Papago Indian Reservation west of Tucson. The Civilian Conservation Corps supplied the labor. At Pueblo Grande the National Park Service assisted a Works Progress expedition, with later support of the CCC. In another realm, a year earlier, in 1938, the founding of an association of museums illustrated two themes characteristic of 1930's Arizona archaeology: teamwork and the campaign to establish the state as a clearinghouse for Southwest archaeology. Practically all of the museums of anthropology in Arizona, New Mexico, Colorado, West Texas, and Southern California formed a loose confederation. Headquartered in the Denver Art Museum, it was appropriately called the Clearing House for Southwest Museum.[4] The extensive role of local government in the Arizona projects ensured a constant barrage of positive headlines about the archaeological efforts. Headlines from press releases exalted the work. One example read: "The Excavation and Repair of a Ruin on the Verde River near Clarksdale, Ariz." The article detailed the project's scientific import, the monies expended, and the labor pools created. Another was more dramatic in its title: "Before Cowhands and Columbus, Out Where the West Began."[5]

New Mexico became the scene of plentiful cooperative archaeological efforts during the 1930s. Scant Works Progress Administration archaeology occurred, however. A conflict of personalities precluded the possibility of greater support from that source. Yet, federal monies entered the state in other forms. Regardless, exchange of research results and a standardization of graduate training came about.

Previous federal work had transpired through the Federal Emergency Relief Administration. The University of New Mexico and Edgar Hewett had accomplished some research during that period. As already mentioned, a cooperative movement had begun with the Jemez Field School in 1935, under the direction of A. O. Bowden of the University of Southern California. Thirty students from over a dozen universities and colleges collaborated with visiting professors. Further dissemination of information began in 1938, when the Department of Anthropology at the University of New Mexico inaugurated publication of a newsletter, *New Mexico Anthropologist*.[6]

Government-funded labor and archaeological work in New Mexico took two forms. In 1938 the National Park Service, in tandem with Civilian Conservation Corps crews in Pueblo Bonito, excavated part of the Chaco Canyon National Monument. Investigations excluding government personnel centered on the University, the Laboratory at Santa Fe, or the Museum of New Mexico. The University itself conducted summer digs and some small Works Progress efforts to excavate and repair several small sites in Chaco Canyon. The National Youth Authority and the CCC also participated in this task. The Museum of New Mexico WPA projects took place at the Pueblos of Abo, Quarai, and Pecos. Later joint ventures with the Works Progress Administration included undertakings at Quarai, Kuaua, and Pueblo Grande in 1940. The Museum of New Mexico and the Arizona State Museum brought to fruition a joint expedition in Las Cruces in 1941.

In this small, sparsely populated state, then, there existed a substantial interest in archaeology. A strong current of cooperation with the existing institutions and between other state agencies pervaded. In essence, the WPA and other federal activities only prompted the expansion of a program that already exhibited definition.[7]

An interesting story, and one that merits elucidation, concerns why New Mexico's survey program fell short of those in other states in the Southwest. Although one had been proposed, implementation never came about. That circumstance stemmed from a personal conflict between Dr. Hewett and the WPA. It revolved around a "controversial" proposal by Dr. Petrullo. During negotiations specific to the drafting of a survey project, Petrullo suggested that the University, the Anthropology Laboratory, and the Museum of New Mexico work together in a blanket federal project designed to incorporate the whole state. As envisioned, a committee would oversee the project, with a director exhibiting high academic standards exercising direct control. Petrullo went on to recommend that the person in charge not come from any of those three institutions. The Laboratory and the University agreed to the plan.[8] Hewett adamantly objected, particularly since he considered himself most qualified to fill the director's position. When Petrullo learned of the altercation, he warned Matthew Stirling at the Bureau of American Ethnology about the impending tempest. In May 1938, Petrullo wrote Stirling, "I have learned that Hewett is leaving for Washington to raise hell with the WPA and the Smithsonian Institution about their basic policies on archaeological projects."[9]

The matter of who became director was of auxiliary importance. The core issue involved instead the monthly reports. The requirement that funded projects submit them was not debated. The information they should contain was! Hewett believed that scientists should be allowed to keep their research findings "secret" for twenty or thirty years. Data would not therefore be made available to other investigators, much less the general public. Undaunted, Hewett submitted yet another proposal. It involved only his museum but did incorporate essentially the same criteria he had previously outlined. Not unexpectedly, neither Petrullo nor the Smithsonian recommended its approval.[10]

This controversy demonstrated at once two things. Strong regional sentiment in archaeology still remained (recall the problems that Matt Stirling faced in Florida during the Civil Works Administration). Also, close professional cooperation persisted between the WPA consultant and the Smithsonian. As a footnote, one might also consider that the National Park Service and Ronald Lee also disapproved of the Hewett format, as did apparently the University.[11]

As the decade neared its end, practitioners in New Mexico and Arizona continued to travel the path toward greater understanding of the region's archaeology, in spite of the absence of significant WPA support. This pattern repeated itself in the East, the South, and the Midwest, only on a larger scale, and principally through the federal Works Administration relief route. This varied course of regional evolution symbolized the status of American archaeology in the Depression. The professionalization process, at times halting and lacking uniformity, depended on a variety of factors for its maturation.

The Midwest experienced extensive archaeological activity with significant cooperative movement. It benefited from the firmly established archaeologic tradition that resided in universities and in historical and archaeological societies. In Nebraska and other midwestern states, the Works Progress relief programs first produced coalescence in that region, then later served to integrate the Midwest and the East.

Two research institutions dominated academic archaeology in Nebraska in the WPA era: the University of Nebraska and the Nebraska State Historical Society and Museum. Located on the same plot of ground, the two planned and otherwise coordinated much of their related research. The University had manifested a strong academic presence in the state for many years previous to the Depression. The State Historical Society worked more closely with the general public through diverse activities and its museum. There, the two institutions encompassed multiple aspects of the discipline's emergent managerial paradigm: the primacy of academic archaeology, a continued importance of museums (both had one), and expanded channels for public dissemination of knowledge.[12]

The Society and the University increased both their function and influence with the inflow of federal money. Additionally, the Smithsonian continued active in this area. The Institution's Waldo Wedel, in 1934 still associated with the University of California, had collaborated with the Society relative to a survey of South-Central Nebraska.[13] The state enjoyed an added bonus—the high level of popular knowledge about the prehistoric past.[14] It was not unusual for farmers and campers to notify the Museum or the University about artifacts or potential sites they spotted. The monthly reports from these two institutions, as well as museum records, indicated when information originated from public disclosure. Based upon this phenomenon, two observations arise. First, investigators of that period felt that the public understood well the "scientific" nature of archaeology. Second, they acknowledged the academic institutional source of professional archaeology.[15] Given the degree of public interest in knowledge about the Indian past, it is not surprising that media coverage of the digs proved effective in stimulating and sustaining support. Information dispersal policies adopted by the Works Administration adroitly supplemented this process. That government body carefully orchestrated press releases and newspaper coverage, resulting in a heightened degree of enthusiasm about archaeology in those areas of the Midwest.[16]

Many talented amateurs operated in these states where interest ran high. They are perhaps best represented by A. T. Hill of Nebraska (a virtual mirror-image of Harold Gladwin in the Southwest). Although not trained formally in archaeology or anthropology, Hill brought to the discipline a consuming interest. Hill had made considerable money in automobile sales. In his spare time, he investigated carefully the Pawnee Indian past. A potent force in Nebraska archaeology, he maintained important positions within the Historical Society for many years and directed several of the expeditions of the 1930s.[17]

The Depression also served to elevate the stature of several professional archaeologists associated with this area. W. D. Strong, who had previously taught at Nebraska, maintained his connections there. He returned several years running with many of his Columbia students. Others arrived as assistant field directors.[18] Two names of note were a product of Nebraska expeditions: Carlyle Smith and Marvin Kivett. From Columbia, where he pursued his doctorate under Strong, Smith then worked in Georgia and in Nebraska, ultimately settling in Kansas for the remainder of his academic career.[19] Kivett as a young man entered the Works Progress Administration system looking for employment. Initially hired as an archaeological laborer, he rapidly moved up to the position of a foreman. After the Depression, he earned degrees in anthropology. Kivett maintained a lifelong relationship with the Nebraska Historical Society, finally becoming its Director.[20]

The Historical Society expeditions, both archaeological and paleontological in scope, took place between June 1936 and September 1941. The paleontological work functioned as a joint effort of the Society and the University. The most spectacular find was the nearly complete skeleton of the largest extant rhinoceros in North America. After processing in its laboratory, the University Museum displayed these specimens.[21]

The archaeological investigations, likewise conducted by both institutions, usually lasted anywhere up to six months. A. T. Hill contributed his time and labor to both the University and the Society. He followed through on extensive summer work done previously in nine counties on Pawnee and protohistoric sites. Village sites and lodges of Pawnee and Upper Republican cultures formed the focus. The fieldwork, quite extensive and covering considerable territory, drew on graduate students, who made up a substantial portion of the supervisorial staff.[22] The University also carried out archaeological work under the direction of Earl H. Bell. He filled the spot that Duncan Strong vacated when he moved, first to the Smithsonian and then on to Columbia. Bell also assimilated graduate students into the operation of the expeditions. Both state agencies provided the necessary laboratory work. For a time, a laboratory was located in the basement of the new capitol building.[23]

Several Nebraska archaeologists participated actively in regional conferences. The Plains Archaeological Conferences proved critical to the creation of a classification for culture manifestations in the northern plains. The paleontological expeditions also stimulated the convocation of similar conferences.[24]

The investigations became important in several other ways. The strong relationship shared with the Smithsonian led to considerable cooperation between the notable archaeological groups. Visitors to these digs included a host of other personages. Even James Griffin made a trip to the Plains and Nebraska. Graduate students received invaluable professional training. The surveys, conducted with the aid of the young scholar Paul Cooper, encompassed several counties at a time. Other participants included Perry Newell and John L. Champe.[25] Champe later went on to begin a long career at the University of Nebraska.[26] The information gained from these expeditions appeared in a variety of journals, such as the publications of the University of Nebraska, the Historical Society, and the Smithsonian.[27]

Institutional forms of cooperation that transcended state boundaries were few, beyond those in the Southeast and the Southwest. Activity in Pennsylvania represented a more typical model of a Works Progress Administration state archaeological project. There, a more substantial university-centered cooperative network dominated. The University of Pennsylvania housed an established anthropology and archaeology program of long-standing repute. This permitted its members to assume the lead in proposing and carrying out projects. Curiously, the University's archaeological inquiries involved research in some areas outside the United States. The Ceramic Repository at the University of Michigan received portions of the materials so obtained. The Works Progress archaeological work undertaken by that institution was unique. Only an eight-project sequence of laboratory work evolved. Thus, white-collar, rather than unskilled, workers benefited from the relief support. In this case, graduate students and the professionals already employed by the University benefited from the relief program.[28]

Formation of an expanded laboratory structure enabled the Department of Anthropology to analyze and catalog a multitude of data from all over the world. The supervisors included Dr.'s Donald Horton, Joseph Berman, and Vladimir J. Fewkes, all associated with the University of Pennsylvania Museum. the reader will recall that Fewkes later accepted a position with the Works Progress Administration in Georgia.[29] Ceramic analysis occupied most of the staff's project time. The enterprise began in October 1935 and terminated in September 1940. Workers achieved tremendous progress in analyzing, preserving, and cataloging material. Specimens from Central America, China, Mesopotamia, Egypt, Haiti, Palestine, Puerto Rico, and Georgia were examined. Supervisors prepared and presented manuscripts for the Conference on Ceramic Technology, held at the University of Michigan in November 1938. Fewkes also initiated compilation of a bibliography on ceramics, construction of a classification system on ceramics, and, with the assistance of W. L. Bliss, the development of a common terminology. The final compilation incorporated more than 35,000 entries.[30]

The remainder of the archaeological expeditions in Pennsylvania comprised field-based projects. These efforts, sponsored by the Pennsylvania Historical Commission, emphasized examination of the state's colonial past rather than its prehistoric heritage. Carried out at the Fort Hill site in short spurts from August 1937 to June 1940, the digs also revealed information about the Algonkian and Monongahela Woodland. Elsewhere in 1939 the Museum and the Society effected a joint inquiry with the aid of the National Youth Authority in Bucks and Montgomery Counties.[31]

Two significant events highlighted the history of New England archaeology in the 1930s. One involved the multistate investigations conducted by the Phillips Academy in Massachusetts. Douglas Byers and Frederick Johnson worked over much of the upper New England area throughout the period. Their research constituted an important step toward erection of a regional interpretation. The geographic size of the small seaboard states lent them easily to this type of approach.[32]

A second noteworthy occurrence stemmed from a university archaeological appointment. Columbia University added W. Duncan Strong to the staff of the Department of Anthropology. Prior to his selection as head of the new Archaeological Laboratory, he had been involved in the Civil Works Administration expedition in California. He had also helped to found academic archaeology at the University of Nebraska.[33] Fully cognizant of the opportunities still available from Works Progress Administration archaeological subsidies, Strong immediately turned this knowledge to his benefit. After his appointment, he visited the Plains WPA projects annually with his students. Strong also worked in areas along the Hudson River and on Long Island. In this manner, several of his students were exposed to a variety of archaeological environments and the scientific problems peculiar to them.[34]

Other states exhibited intrastate cooperation between institutions. In Wisconsin, the state's Archaeological Society and the Madison Park Board excavated and repaired mounds in and around Madison. The City of Milwaukee sponsored similar restoration efforts.[35] On a larger scale, the Wisconsin Archaeological Society and the State Historical Museum collaborated to conduct surveys and limited expeditions beginning in 1934.[36] Cooperation also occurred in this manner between the University of Wisconsin and the Milwaukee Public Museum.[37]

In Indiana, the Historical Society commenced work in 1934.[38] The Society and the State Historical Bureau cosponsored a Works Progress archaeological project in Vanderburgh and Warrick Counties. They focused attention on the Angel Mounds, which gave up over 250,000 specimens. According to the supervisor, Glenn A. Black, some of the shards closely exhibited traits similar to the cultures of the Southeast.[39]

The University of Chicago constituted the most dynamic element within this region, with Illinois as the primary site for the fieldwork of Fay-Cooper Coles's Department of Anthropology. The University had previously set up a tree-ring dating project for eastern archaeological study. Donation of a building and the availability of a research grant brought Florence Hawley there to teach and research for three years.[40] The Illinois State Museum, also active, sponsored a Works Progress statewide museum project at Cahokia and at Fort Massac.[41] Illinois also hosted the conference of the Mississippi Valley Federation of Archaeological Societies. First convened in October 1939, it included representatives from Illinois, Missouri, Kentucky, Indiana, Kansas, and California.[42] By 1940, Thorne Deuel, in collaboration with Winslow Walker, expanded the Illinois relief museum extension project into areas threatened by highway construction.[43] The Museum and the University also pursued other joint operations under the Works agency. Another University of Chicago excavation in 1940 occurred at the Kincaid site, with assistance offered by students and a WPA relief labor force. Visiting archaeologists included Frank Setzler of the Smithsonian and W. C. McKern, co-author of the Midwestern Taxonomic System. The information revealed helped to provide clues about the archeological sequence within the Tennessee–Cumberland component. The newspaper coverage of all this activity stated results in a positive manner: "Archaeological Research Reveals Illinois History."[44]

Further west, the Montana School of Mines and the University of Montana organized two major digs funded for over $80,000. Multiple sponsors included Montana State University, Eastern Montana Normal School, and the Montana State Highway Commission. These expeditions, supervised by H. M. Sayre and William T. Mulloy, Jr., continued sporadically from July 1937 until December 1941. Workers unearthed a good deal of information and well over 100,000 artifacts. These were later cataloged, with many recorded pictographs. The evidence established not only that a semi-settled agricultural people had inhabited eastern Montana but that pottery makers had been present in the west and central portions of the state.[45] The investigation that transpired in Montana, initiated before New Deal relief became available, thus drew upon an established archeological network. The schools, which published the results of the digs, arranged for graduate student participation. Newspaper coverage for this archaeological episode continued as late as August 1941. Headlines touted the importance of the "WPA Survey of Indian Caves."[46]

A similar situation existed in the Rocky Mountains, where Wyoming followed a course of development similar to Montana's. Again, a college furnished the headquarters for these projects. Archaeological study became further entrenched into the web of a university structure.

The projects, among the most extensive undertaken in the Pacific Mountain region, lasted from December 1938 to December 1940. Total costs were in excess of $200,000, with additional sponsor funds obtained from the Dubois Chamber of Commerce and the Natrona County Board of Supervisors. A high percentage of the expenses went for laboratory and photography enterprises, by which middle-class workers shared in the relief program. Even women found opportunity, most often as clerical workers, but on occasion in professional positions as well.

Robert B. David and Ted C. Sowers supervised the projects, assisted by graduate students much of the time.[47] The rapidly changing climatic conditions in the Rocky Mountains precluded lengthy field expeditions. Personnel labored often at altitudes of between 9,000 and 11,000 feet. Since a laboratory processing unit existed, supervisors directed much greater emphasis to this facet of research than had other state projects. They channeled several thousand recovered artifacts and petroglyphs through the laboratory at Laramie University. As had been done in Montana, considerable effort involved preserving and cataloging petroglyphs. Coordinated with the Sundance Museum in Natrona County, which supplied many of their resources, inquiry revealed predominately Shoshoni burials and artifacts. Cave work on Medicine Creek, Crook County, 61 miles northwest of Sundance, divulged petroglyphs and the remains of the Plateau Indians, parts of which were analyzed at the Casper laboratory.[48]

These undertakings were comparatively extensive, given the size of the region. The supply of information, important for the archaeological community, also encouraged substantial public interest as well. The survey, combined with the recognized importance of this work, became critical ingredients in the movement to construct a museum to house the newly found artifacts and copies of the petroglyphs, according to consultant Vincenzo Petrullo.[49]

A third level of relief work involved individual institutions. Although pursued independently, this work nonetheless contributed to the growing prestige of academic archaeology and the expansion of knowledge about the American prehistoric Indian.

Neither California nor Nevada experienced extensive or well-publicized, federally sponsored archaeological fieldwork. Relief archaeology did not alter this circumstance. The University of California and its Museum did conduct a paleontological expedition, accomplished with the aid of WPA labor. The work was later incorporated into a larger bibliography on vertebrate paleontology.[50] Elsewhere, the Southwest Museum in Los Angeles, the Santa Barbara Natural History Museum, and the San Diego Natural History Museum continued to explore constantly throughout the relief period in the San Joaquin and Sacramento Valleys, the Tehachapi Mountains, the mountains of San Diego and Nevada, and the Mohave.[51]

One last facet of relief archaeological projects in California concerned the sponsorship and the dearth of publicity given the expeditions. John W. Winterbourne supervised two digs, one subsidized by the Board of Education for Santa Ana City Schools, the other by the Fullerton Union High School District and Junior College. These projects operated intermittently between May 1937 and December 1938. The areas excavated focused on the Irvine Ranch and the upper Newport Bay district. The information garnered supported the earlier Kroeber generalization that members of the Shoshonean family had inhabited this area.[52]

The Northwest, which witnessed considerably more activity than California, saw each state thrive at a different level of institutional development. Collectively, the work undertaken in Washington and Oregon produced several results. Labor support derived from several agencies prompted the expansion of knowledge of several regions. Oregon and Washington collaborated with the Smithsonian Institution, the Civilian Conservation Corps, and the National Youth Authority.

The story of Oregon archaeology began in 1933, when L. S. Cressman arrived on the Eugene campus. His appointment held special significance. It was the first university-based position in archaeology in the Pacific Northwest. Funding, obtained from the Carnegie Institution, the U.S. Bureau of

Reclamation, and, later, the U.S. Geological Survey, made possible Cressman's research. The National Youth Authority in spring 1940 assisted in the Willamette Valley expedition, organized by the Willamette University Anthropology Department and Museum.[53]

Archaeological work in Washington State centered around the building of the Grand Coulee Dam in central Washington and the ongoing investigation of the Columbia Basin. The Chief Engineer of the dam supported a program that included a survey and digging of test sites. Initially begun by Alex Krieger under the sponsorship of the Spokane and Eastern Bank, Phillip Drucker took over when Krieger moved to the University of Texas. The National Youth Authority furnished the labor for their expeditions. Additional support came from the Department of Anthropology at the University of Washington and from Columbia University. This program also exhibited a high degree of internally generated state cooperation. Besides the University of Washington, the Spokane Museum and Eastern Washington State Historical Society participated.[54]

New Deal work in Utah and Colorado relied on non–Works Administration sources. The National Park Service carried out extensive surveys in Zion National Park as part of an ongoing survey in all national parks. The University of Utah also conducted a smaller survey in Nine Mile Canyon in Carbon County. The Yale Survey, led by Helmut de Terra and supported by the Carnegie Institution, assessed several High Plains sites in Colorado throughout the decade. Frank H. H. Roberts instigated research for the Bureau of American Ethnology near Fort Collins in 1934. Waldo Wedel headed an expedition into Colorado in 1938 for the U.S. National Museum. The Field Museum from Chicago and Harvard's Peabody also subsidized researchers there at the same time. One interesting side-note associated with Colorado archaeology involved the printing of a leaflet by H. M. Wormington of the Colorado Museum of National History. Available at a cost of ten cents, "The Amateur Archaeologist" represented an attempt to enlighten the layman about scientific techniques of investigation in archaeology. Multiple distributions of this documented occurred.[55]

Considerable amounts of cooperation within each state transpired. The collaborations developed due to the actions of professionals who did not necessarily share interests in common cultural areas. Other positive thrusts persisted as well. First, the movement to elevate the general status of archaeology and then the centrality of the university-based and professional discipline retained its impetus. The campaign to control the activity of the amateur element endured, as it did in other states.

The work carried out in Texas and Oklahoma, both widespread and lengthy, rivaled that of the more well-publicized southeastern programs. In terms of the number of jobless people employed and the amount of information recovered, the two regions were approximately equivalent. The Southwestern Works Progress Administration expeditions of the two states were parallel in several respects. The indigenous dry climates aided in the preservation of artifacts and bones. Considerable paleontological research occurred under the Works Agency, stimulating further inquiry after World War II. Digs in both states, which encompassed several counties, received favorable attention in the media. One prominent example pronounced: "Federal Grants Help Diggers to Uncover the Story of Yesterday."[56]

Texas hosted six Works Progress projects, which benefited from funds allotted in excess of $200,000. In West Texas, Floyd V. Studer at the West Texas State Teachers College spearheaded investigations from February 1938 to July 1941. Recovered were more than 20,000 shards and 20,000 artifacts of the Pan Handle Culture.[57] The University of Texas sponsored a program, from August 1938 to December 1940, that covered huge portions of the state. A. T. Jackson supervised twenty-four county surveys and numerous sites. Portions of the information, especially that regarding the many projectile points unearthed, appeared in the *University of Texas Anthropological Papers* and the *Bulletin of the Texas Archaeological and Paleontological Society*.[58]

Two other schools in the Lone Star State directed relief archaeological programs. Texas Technological College managed a dig in Jones County between October 1939 and March 1941. Joe Ben Wheat supervised the research into village sites and middens. Sul Ross College undertook a small project in Presidio County between July 1938 and June 1939, where J. C. Kelly continued his investigation into what he called the "Bravo Valley Aspect." These took place at the village sites and mounds of the Millington Site, situated some five miles above Presidio, Texas. In all these instances, efforts followed a similar pattern: if only scant cooperation among local schools existed, none emanated from outside the region.[59]

Related paleontological activity again centered in the West Texas State Teachers College and the University of Texas. the reader will recall that archaeology and paleontology, grouped together under the Works Progress Administration, often shared resources and pooled results. First with C. Stuart

Johnston in command, then under Wayne G. Christian, and finally led by William F. Read, West Texas personnel performed surveys and excavations between March 1939 and March 1941. They obtained specimens mostly from the Upper Pliocene. The work revealed both insect remains and those of large and small animals. The latter group included rhinoceros, dogs, antelopes, horses, turtles, tiger cats, sloths, camels, deer, elephants, rodents, and even a mastodon.[60]

The University of Texas, with E. H. Sellards in charge, ran a paleontological reconnaissance from October 1939 to March 1941. The survey and excavations of twenty-one counties recovered a tremendous amount of fossil material from the Permian, Triassic, Cretaceous, Pliocene, and Pleistocene epochs. Materials unearthed encompassed meteorites, tens of thousands of remains, and several complete skeletons of dinosaurs. Finally, the Texas College of Mines and Metallurgy organized a small project under the direction of Raymond Miller. Operated from July 1938 to January 1939 in Brewster County, it uncovered over five hundred specimens.[61]

The University of Oklahoma stood out as the only Oklahoma state institution that conducted Works Progress Administration archaeological research. Archaeological and paleontological projects took place from June 1935 until September 1941. As was the case in Texas, the information recovered not only added greatly to the knowledge of the area, it also further established the position of the University as the essential scientific center for this type of inquiry. The state press consistently publicized the activity surrounding the Oklahoma research. Little in the way of joint enterprises with any out-of-state institutions resulted, however.

Forrest E. Clements and his students supervised the Oklahoma digs for the entire WPA period. Already firmly ensconced in state professional circles, Clements aggressively utilized relief archeology in eleven counties for all its worth. The Spiro Mound explorations in LeFlore County, where some of the sites revealed thousands of burials, was a cooperative effort. It included the University of Tulsa and a third party, a laboratory in Cleveland County, which processed the thousands of shards and artifacts. Funding alone for this long-term project exceeded $150,000.[62] The actions here were covered in several articles. One press release, in particular, proclaimed: "Archaeology in Oklahoma was aroused. Panhandle investigations have resulted in 32 scientists to organize Oklahoma Archaeology Society which worked to promote approval of $83,000 WPA project."[63]

The accompanying paleontology program, supervised by J. Willis Stovall, embraced research in eleven counties from December 1937 to March 1942. Through a series of small projects, Stovall directed considerable attention to unearthing Pliocene, Jurassic, and Permian reptiles. The nearly complete remains of a brontosaurus and a diplodocus were among the most noteworthy findings. The Oklahoma State Archaeological Society published portions of this research.

An interesting piece in the Forth Worth (Texas) *Star Telegram* reported WPA activity in Oklahoma. It revealed some of the local color attached to these types of relief efforts. The column heading read: "WPA Workers Digging in Indian Mounds in Oklahoma Scoff of Rumors of "Ha'nts." Workers interviewed were quoted as saying, "Shucks, there ain't ha'nts in these diggings. That's a lotta hooey."[64]

There was one instance in which Oklahoma archaeologists' proved willing to participate in a cooperative effort on a national scale. It occurred outside the aegis of the relief operations, but nonetheless reaped dividends. The activities of the State Surveys Committee in 1933 provided a partial stimulus to this latter activity. The Committee alerted the University of Oklahoma that items from the Spiro Mound periodically came on the artifact market in the East. The eventual result of this communication took the form of a state law. Thereafter, it became illegal for an individual to sell artifacts unless he was associated with a scientific organization.[65]

Some of the most extensive Works Progress archaeological activity in the East centered in New Jersey, although the expeditions involved neither a university nor were they cooperative in scope. Yet, the work sponsored by the New Jersey State Museum certainly increased the importance of this institution. Large amounts of materials proved fruitful in building more thorough temporal and interpretive systems. As in Texas, the projects investigated vast portions of the state, specifically ten counties.

Work in New Jersey spanned the period from April 1936 to March 1941. The participation of many urban dwellers was evident. Overall quality of the archaeology accomplished has been graded as, at best, mediocre. The Archaeological and the Historical Societies of New Jersey published many of the reports made about the sites. The sponsors printed and distributed nineteen pamphlets concerning the history of the Indian culture groups investigated. Displayed in the Museum were many of the hundreds of thousands of preserved artifacts. Complementary efforts also included the creation of a file that abstracted the materials garnered. The file proved more fruitful for use by academic archaeologists. The state agency of the Federal Writer's Project even participated in this program by preparing a series of bulletins about the finds. The effect of the archaeological labors on the State Museum proved to be the most significant consequence of the New Jersey programs. They constituted a windfall since prior to 1934 archaeological investigations by that agency had been nonexistent. Press coverage here also linked archaeology with other sciences. Simultaneously, the publicity reinforced the perceived importance of the relief work. One news item from 1940 read: "Footprints 100,000 years old," and "[Work for] unemployed technicians and clerical workers have helped to carry on valuable research in archaeology, paleontology, physics, chemistry and other branches of science."[66]

The remainder of New England relief archaeology operated through the various state societies and museums. To cite one example, the Museum of the American Indian in Vermont researched near Fort Ticonderoga.[67] In Delaware, the Archaeological Society of Delaware carried out small excavations. Similar activities in Rhode Island, Connecticut, New Hampshire, and Maine were repeated.[68] New York saw the activist Rochester Museum of Arts and Sciences initiate work throughout the state.[69]

In the Midwest, archaeological tradition in Ohio revolved around the Ohio State Archaeological and Historical Society. The Society's Museum received Works Progress Administration money to prepare a bibliography of American archaeology. Eight workers compiled this document, which emphasized research in the Mississippi Valley.[70] The Department of Archaeology at the Ohio State Museum, also a participant in state archaeological activity, researched Fort Ancient Culture. One unique project in Ohio attracted considerable attention. Work on an Indian snake edifice was linked with tourism. Headlines declared in 1938 that "Ohio Serpent Mound Attracts Thousands of Tourists." The article went on to declare how $80,000 worth of improvement was devoted to a serpent effigy.[71]

More midwestern Works Progress Administration efforts centered in Missouri. There, the Academy of Science of St. Louis accomplished considerable research. The basic goal of these operations concentrated on the collection of material to form a nucleus for a scientific museum in St. Louis. On occasion, outside agencies ran expeditions. The state also witnessed organization building activity in the profession. The Academy digs, noteworthy for both their longevity and the brevity used in communicating research findings, produced reports filed with machinelike regularity. Little but the most general of information resulted, though. Robert McCormick Adams supervised projects in five counties from December 1939 until April 1942.[72] His monthly reports systematically failed to specify the publication status of the work, nor did they offer evidence of collaboration with any other institution.[73]

Missouri's efforts in the realm of archaeological preservation achieved a higher character, with the founding of the Missouri Archaeological Society in 1934. It immediately commenced work on a state survey.[74] The University of Missouri cooperated on a small scale with various agencies throughout the decade, including the FERA and the National Youth Authority.[75] Under its auspices, Professors Berry, Wrench, Chapman, and Mack surveyed several sites and conducted digs. The Smithsonian, through Waldo Wedel of the U.S. National Museum, in 1938 sponsored excavations in Missouri, as it did in Nebraska.[76] The state financed parties in eastern Missouri led by Paul Munger and Winslow Walker, the latter one of the most well-traveled archaeologists of the decade.[77]

Farther north on the Plains, North and South Dakota saw programs smaller but more coordinated than those in Missouri. The North Dakota Historical Society in 1934 first excavated on Mandan Indian sites, using Civilian Conservation Corps labor. Later, after 1935, and on a limited and occasional basis, the Society produced a survey of Missouri Valley pottery, using WPA labor.[78] South Dakota, through its University, became more active than its sister state, despite not then having a program in anthropology. Elmer E. Meleen, a self-educated amateur, acted as the supervisor in most operations between June and December 1939. Data from material from five counties, analyzed in the University laboratory at Vermillion, later found its way into print.[79] The University of South Dakota and Columbia University also pooled resources in one operation between July and September 1939. A. C. Spaulding supervised the surveys and excavations.[80]

Minnesota's minor program efforts followed the format of South Dakota. School districts and county commissioners fostered several local projects, usually associated with a park.[81] The Anthropology Department of the University of Minnesota developed three seasons of work between June 1939 and October 1941. L. A. Wilford supervised these undertakings in nine counties. The projects uncovered and returned to the University for analysis considerable material associated with Woodland and Mississippi Cultures.[82] The University and the Minnesota Historical Society inaugurated a joint operation for a statewide survey that included archaeological and historical sites. The Society supported in August 1941 one small dig in Big Stone County at the Big Stone Lake Burial Mound. A last cooperative effort occurred between the Minnesota Division of State Parks and the National Park Service. Begun in November 1936 at Fort Ridgely, the dig sought information on the Sioux outbreak of 1862.[83]

Limited archaeological activity appeared in the two remaining Plains states, Iowa and Kansas. The State Historical Society of Iowa, first joining forces with the Federal Emergency Relief Administration in June 1935, accumulated information on a small scale until September 1938.[84] Kansas's principal archaeological effort derived from Smithsonian investigations that included Kansas as a part of a larger survey led by Waldo Wedel. This particular program for the U.S. National Museum began in 1937 near Kansas City and Doniphan.[85] A positive move forward for academic archaeology occurred with the full-time appointment of Loren C. Eisley as Professor of Anthropology at the University of Kansas. He started expeditions the following summer in 1937.[86]

The Midwest symbolized an archaeological region in transition at all levels, primarily dependent for its subsequent development on the founding of universities and the continued support from state agencies. The interactive nature of federally assisted investigations further highlighted the professional nature of the work. It also aided in the rapid dissemination of the information.

New Deal archaeology in the South, beyond those areas already discussed, was confined to a few small projects divided between Mississippi, Florida, Arkansas, and North Carolina. Importantly, activity at these sites usually involved known archaeologists. Moreover, the interaction of a variety of agencies documents that institutional developments there paralleled those of other states, despite the lack of large-scale relief programs. Mississippi initiated fieldwork with the State Archaeological Survey, overseen by the Department of Archives and History and the Smithsonian. Moreau B. Chambers and two students excavated sites in Clay and Oktibbeha Counties. Their labors continued later on with FERA funds and Civilian Conservation Corps workers. Albert Spaulding oversaw excavations of Chickasaw sites near Tupelo under the Works Progress Administration.[87] Other surveys took place under Jesse Jennings, who classified additional Chickasaw areas in Jefferson and Lee Counties.[88] The digs in Florida proved to be of high value. Workers unearthed very large human bones at a depth which placed Indians at the site before the Spanish invasion. In the summer of 1940, a Columbia University field party commenced a survey in West Florida. Gordon R. Willey and Richard Woodbury supervised excavation of seventy sites, six of them stratigraphically. Later, the Florida Archaeological Survey backed a survey in Hillsborough County.[89]

North Carolina and Arkansas carried out a degree of inquiry during this time. North Carolina had the University of North Carolina and the North Carolina Archaeological Society to stimulate research. The Society employed Joffre Coe to examine a village site in the Piedmont area on Caraway Creek.[90] Later, Coe and Robert Wauchope surveyed under University auspices Montgomery and Orange Counties, along with work on three specific sites: the Frutchey Mound, the Frutchey village site, and the Occaneechi Village site. This activity followed work undertaken at these sites earlier by the Department of Conservation in 1937.[91]

Arkansas archaeology flourished through the University of Arkansas, coupled with the enthusiasm of S. C. Dellinger. Already active by 1934, Dellinger supervised several small projects in village sites in Hot Springs, Garland, and Union Counties between 1937 and 1940. Collections included human and dog burials, houses, and pottery showing affinities to the Cole Creek Complex in Louisiana. The Union County operations, primarily zoological in nature, were also subjected to laboratory analysis.[92] *American Antiquity* later published the information from these digs.[93] Henry Roberts of the Carnegie Institution directed other state activities.[94]

The last New Deal archaeological work appeared in Puerto Rico in 1938 as part of a larger survey. It resulted in an index of all events pertaining to the history of the island.[95]

A final category meriting attention centers on the unique status of the University of Michigan. Ann Arbor made a primary contribution to 1930's relief archaeology, essentially in the form of laboratory and analytical investigations. The campus did not engage in relief fieldwork, but it stood as the best example of what could be accomplished through cooperative analysis. The Repository and the Museum of Anthropology completed extensive research, as tremendous amounts of material from the southern digs came there, as was also the case for artifacts from the Near East. Carl Guthe had labored heartily to create such an institution. He succeeded.[96]

Works Progress Administration projects manifested one last dimension. They attracted the attention of the press and national media in places other than those already described. Ohio and South Dakota reaped similar coverage to that generated by the previously noted Georgia Civil Works Administration efforts. One Works Agency press release discussed the thousands of tourists attracted to the Ohio Serpent Mound. Some $80,000 worth of improvements were made at the mound site to accommodate the persistent influx of people. These included a home for the supervisor, a garage, a main building, a shelter house, a storage building, parking areas, picnic tables, and ground storage.[97] A similar announcement concerned the Rapid City, South Dakota, Dinosaur Park, where $11,845 in federal money expanded and improved the site.[98] The Washington WPA office publicized North Dakota's Indian work. Similar entries appeared in the local papers.[99] Nebraska's press quickly printed articles about that state's Historical Society.[100] The accelerated activity reported helped to disseminate the message about the "scientific" importance of archaeology. Two major themes that received constant coverage in the press focused on the beneficial economic effects of relief digs and the "scientific" nature of the work. The widely read newspapers probably did more to communicate effectively the latter theme than any of the many pamphlets printed to educate the amateur archaeologists and the public alike.

Notes

INTRODUCTION

1. Lawrence R. Vesey, *The Emergence of the American University* (Chicago: The University of Chicago Press, 1965), 1–17; Daniel Kevles, *The Physicists: The History of a Scientific Community in Modern America* (New York: Alfred A. Knopf, 1978), 1–7; Robert H. Wiebe, *The Search for Order, 1877–1920* (New York: Hill and Wang, 1967); Bernard Barber and Walter Hirsch, eds., *The Sociology of Science* (New York: Free Press of Glencoe, 1962); and John Madge, *Origins of Scientific Sociology* (New York: Free Press of Glencoe, 1962).

2. J. O. Brew, ed., *One Hundred Years of Anthropology* (Cambridge: Harvard University Press, 1968), 5–29; and Gordon R. Willey and Jeremy A. Sabloff, *A History of American Archaeology* (New York: W. H. Freeman & Co., 1993), 3rd ed. rev., 1–11.

3. Several meaningful articles calling for broad, interpretive surveys have influenced this effort. Among them are: William E. Leuchtenburg, "The Pertinence of Political History: Reflections on the Significance of the State in America," *The Journal of American History* 73 (1986): 585–600; Thomas Bender, "Wholes and Parts: Continuing the Conversation," *The Journal of American History* 74 (1987): 123–30; William H. McNeill, "Mythistory, or Truth, Myth, History, and Historians," *American Historical Review* 91 (1986): 1–10; and Theodore S. Hamerow, *Reflections on History and Historians* (Madison: University of Wisconsin Press, 1987).

4. Willey and Sabloff, *American Archaeology*, 1–10; Walter W. Taylor, *A Study of Archaeology*, (Carbondale: Southern Illinois University Press, 1964), 9–23; and Sol Tax, "The Setting of the Science of Man," in Sol Tax, ed., *Horizons of Anthropology* (Chicago: Aldine Pub. Co., 1964), 15–20. The nature of the older documentation, those cultures that did not leave any pottery, burial procedures or the like, also preclude any implicit interpretation from the data, as Thomas Kuhn suggests. These obstacles, coupled with institutional-oriented problems, make it unlikely to some of its practitioners that a widely accepted ontological paradigm will be developed. Letter, Frederick Johnson to author (7 December 1983); and Thomas Kuhn, *The Structure of Scientific Revolutions*, 2nd ed., enl. (Chicago: University of Chicago Press, 1970), 10–11, 13–15, 35–42, 160–62, 175–76. See also Lewis R. Binford and Jeremy A. Sabloff, "Paradigms, Systematics, and Archaeology," *Journal of Anthropological Research* 38 (1982): 137–53, for a conclusion that archaeology as a science is more appropriate, and an opposing view in R. C. Dunnel, "Science, Social Science, and Common Sense: The Agonizing Dilemma of Modern Archaeology," *Journal of Anthropological Research* 38 (1982) : 1–25.

5. A. Irving Hallowell, "The Beginnings of Anthropology in America," in *Selected Papers from the American Anthropologist: 1880–1920*, ed. Frederica De Laguna (Evanston, IL: Row, Peterson and Co., 1960), 1–90; and T. K. Penniman, 3rd ed., rev., *A Hundred Years of Anthropology* (London: Gerald Duckworth & Co., 1965), 321–60.

6. Talcott Parson, "Professions," in *The International Encyclopedia of the Social Sciences* 12 (1968): 536–46; and see also the following for a further discussion of professionalization, sociology of science, and institutionalization of society: John A. Garraty, *The New Commonwealth: 1877–1890* (New York: Harper & Row, Pub., 1968); Haskell, *Professional Social Science*, 1–47; Wilbert E. Moore, *The Professions: Roles and Rules* (New York: Russell Sage Foundation, 1970); Talcott Parsons, "The Professions and Social Structure," *Social Forces* 17 (1939): 457–67; Everett C. Hughes, "Professions," *Daedalus* 92 (1963): 655–68; Bernard Barber, "Some Problems in the Sociology of the Professions," *Daedalus* 92 (1963): 669–88; John J. Beer and W. David Lewis, "Aspects of the Professionalization of Science," *Daedalus* 92 (1963): 764–84; George H. Daniels, "The Process of Professionalization in American Science: The Emergent Period, 1820–1860," *Isis* 58 (1967): 151–66; David Edge, "Is there too much sociology of science?," *Isis* 74 (1983): 250–56; Michael J. Mulkay, "Methodology in the sociology of science: Some reflections on the study of radio astronomy," *Social Science Information* 13 (1974): 107–19; Michael J. Mulkay, "Action and Belief or Scientific Discourse? A possible way of ending intellectual vassalage in social studies of science," *Philosophy of the Social Sciences* 2 (1981): 163–71; Michael J. Mulkay and G. Nigel Gilbert, "Putting Philosophy to Work: Karl Popper's Influence on Scientific Practice," *Philosophy of the Social Sciences* 2 (1981): 389–407; David O. Edge and Michael J. Mulkay, *Astronomy Transformed: The Emergence of Radio Astronomy in Britain* (New York: John Wiley & Sons, 1976); Lewis E. Auerback, "Scientists in the New Deal: A Pre-War Episode in the Relations between Science and Government in the United States, *Minerva* 3 (1965): 457–82; Joseph Ben-David and Awraham Zloczowee, "The Scientific Community in American and European Universities," *European Journal of Sociology* 3 (1960): 45–84; Stephen P. Turner, "Interpretive Charity, Durkheim, and the 'Strong Program' in the Sociology of Science," *Philosophy of the Social Sciences* 2 (1981): 231–43; Merrilee H. Salmon, "Ascribing Functions to Archaeological Objects," *Philosophy of the Social Sciences* 2 (1981): 19–26; and Elvi Whittaker, "Anthropological Ethics, Fieldwork and Epistemological Disjunctures," *Philosophy of the Social Sciences* 2 (1981): 437–51; Larry Laudan, "The Pseudo-Science of Science?," *Philosophy of the Social Sciences* 2 (1981): 173–98; Bernard Berelson, *Graduate Education in the United States* (New York: McGraw-Hill Book Co., Inc., 1962); William C. Devan, *Higher Education in Twentieth Century America* (Cambridge: Harvard University Press, 1965); and Michael J. Storr, *The Beginning of Graduate Education in America* (Chicago: University of Chicago Press, 1953).

7. See Richard Whitely, "Umbrella and Polytheistic Scientific Disciplines," *Social Studies of Science* 6 (1976): 471–97; Herbert Mentzel, "Planning the Consequences of Unplanned Action in Scientific Communication," in *Communication in Science: Documentation and Automation*, 33–69, ed. Anthony De Reuck and Julie Knight (New York: Little Brown & Co., 1967); Herbert Mentzel, *The Flow of Information Among Scientists*, 2 vols. (New York: Columbia University Press, 1958); A. J. Meadows, *Communication in Science* (Stoneham, MA: Butterworth, 1974); C. E. Nelson and D. K. Pollock, eds., *Communication Among Scientists and Engineers* (Lexington, MA: D. C. Heath, 1970); Thomas F. Gieryn and Robert K. Merton, "The Sociological Study of Scientific Specialties," *Social*

Studies of Science 8 (1978): 257–61; and Belver C. Griffith, Marilyn J. Jahn, and A. James Miller, "Informal Contacts in Science: A Probabalistic Model for Communication," *Science* 173 (1971): 164–66.

8. See the *Comprehensive Dissertation Index 1861–1972, Social Sciences*, Vol. 17 (Ann Arbor: Xerox University Microfilms, 1973), 1–190. Between 1895 and 1950 twenty-two institutions granted 476 degrees. Six institutions dominated. They were, in descending order: Harvard—99; Columbia—70; UC Berkeley—57; Chicago—55; Yale—46; and Penn—31. Average graduates for each year, then, is rounded off to 9. The majority of degrees began to appear in the late 1920s and increased steadily throughout the 1930s and 1940s. An average of 9 graduates a year would not easily fill all the necessary positions available in a large country.

9. Warren O. Hagstrom, *The Scientific Community* (New York: Basic Books, 1965), 23–42, 69–85, 105–54, 168–87, 202–208, 216–18, 221–26, 273–85; Haskell, *Professional Social Science*, 19; Daniel J. Amick, "An Index of Scientific Elitism and the Scientist," *Social Studies of Science* 4 (1974): 1–16; and Noah E. Friedkin, "University Social Structure and Social Networks Among Scientists," *American Journal of Sociology* 83 (1978): 1445–65.

10. Haskell, *Professional Social Science*, 136–37.

11. Quimby, "History of WPA Archaeology," 110–23.

PROLOGUE

1. Curtis M. Hinsley, Jr., *Savages and Scientists: The Smithsonian Institution and the Development of American Anthropology, 1846–1910* (Washington, DC: Smithsonian Institution Press, 1981), 1–12; and Willey and Sabloff, *American Archaeology*, 38–44.

2. Nathan Reingold, "Definitions and Speculations: The Professionalization of Science in America in the Nineteenth Century," in *The Pursuit Of Knowledge In The Early American Republic: American Scientific and Learned Societies from Colonial Times to the Civil War*, ed. Alexandra Oleson and Sanborn C. Brown (Baltimore: Johns Hopkins Press, 1976), 50; and Francis Bacon, *The Great Instauration and New Atlantis*, ed. J. Weinberger (Arlington Heights, IL: AHM Pub. Corp., 1980), 79–80. See also Curtis M. Hinsley, Jr., "Revising and Revisioning the History of Archaeology: Reflections on Region and Context," in *Tracing Archaeology's Past: The Historiography of Archaeology*, ed. Andrew L. Christenson (Carbondale: Southern Illinois University Press, 1989), 80–96, for an excellent discussion of regionalism.

3. Willey and Sabloff, *American Archaeology*, 24–36; David S. Brose, "The Northeastern United States," in *The Development of North American Archaeology*, ed. James E. Fitting (Garden City, NJ: Anchor Books, 1973), 84–116; James B. Stoltman, "The Southeastern United States," in Fitting, *North American Archaeology*, 117–50; Jesse E. Jennings, *The Prehistory of North America* (New York: McGraw-Hill, 1968), i; and William H. Goetzman, *Army Exploration in the American West, 1803–1863* (New Haven: Yale University Press, 1959), and Reingold, "Definitions and Speculations," 33–69. Reingold categorizes the scientific community as comprising researchers, practitioners, and cultivators. For another example, see John Finley Freeman, "Religion and Personality in the Anthropology of Henry Schoolcraft," *Journal of the History of the Behavioral Sciences* 1 (1965): 301–13.

4. Kuhn, *Scientific Revolutions*, 15.
5. Ibid.
6. Nineteenth-century naturalists attempted to explain various phenomena by means of simple identification of causation; questions about the more basic reasons that underlay the operation of natural forces were generally ignored. Naturalism, by virtue of its approach, remained closely associated with empiricism. See Fred W. Voget, "Progress, Science, History and Evolution in Eighteenth- and Nineteenth-Century Anthropology," *Journal of the History of the Behavioral Sciences* 3 (1967): 132–55.
7. By the 1840s scientists received instruction in other areas in the colleges and universities. The specific reference here pertains to archaeologists and anthropologists. For further delineation, see Stanley M. Guralnik, "The American Scientist in Higher Education, 1820–1910," in *The Sciences in the American Context: New Perspectives*, ed. Nathan Reingold (Washington, DC: Smithsonian Institution Press, 1979), 99–142. Baconian empiricism rested on two precepts: first, technology changed the course of history; second, to interpret nature one relied on induction. Bacon also emphasized that the primary task of the scientist was not to theorize, but rather to accumulate sufficient amounts of data. As such, scientific knowledge derived from first-hand laboratory or field experience. See Paolo Rossi, "Baconianism," in *Dictionary of the History of Ideas: Studies of Selected Ideas* 1 (1973): 174–75, and Bacon, *New Atlantis*, 79. See also Kevles, *The Physicists*, 1–12, 37; and Reingold, "Definitions and Speculations," 38–39.
8. Kuhn, *Scientific Revolutions*, 16; and Franz Boas, "The History of Anthropology," in *Readings In The History Of Anthropology*, ed. Regna Darnell (New York: Harper & Row, Pubs., 1974), 260–73.
9. Hinsley, *Savages and Scientists*, 17–29; A. Hunter Dupree, *Science in the Federal Government: A History Of Policies and Activities to 1940* (Cambridge: Harvard University Press, 1957), 66–83; Nathan Reingold, "Alexander Dallas Bache," in *Dictionary of Scientific Biography*, Vol. 12 (1970): 203–205; Nathan Reingold, "Alexander Dallas Bache: Science and Technology in the American Idiom," *Technology and Culture* 11 (1970), 57; Robert Post, "Science, Public Policy, and Popular Precepts: Alexander Dallas Bache and Alfred Beach as Symbolic Adversaries," in Reingold, *Sciences in the American Context*, 77–98; Nathan Reingold, "Joseph Henry," in *Dictionary of Scientific Biography*, Vol. 6 (1970): 277–81; Wilcomb E. Washburn, "Joseph Henry's Conception of the Purpose of the Smithsonian Institution," in *A Cabinet of Curiosities*, ed. Walter Muir Whitehill (Charlottesville, VA: University of Virginia Press, 1967), 106–66; and Nathan Reingold, ed., *Science in Nineteenth Century America, A Documentary History* (New York: Hill and Wang, 1964), 59–107.
10. Thomas L. Haskell, *The Emergence of Professional Social Science: The American Social Science Association and the Nineteenth-Century Crisis of Authority* (Chicago: University of Illinois Press, 1977), 66; and William Stanton, *The Leopard's Spots: Scientific Attitudes Toward Race in America, 1815–59* (Chicago: University of Chicago Press, 1960), 65–88. The polyethnic controversy, closely debated in private organizations like the American Ethnological Society as well as in larger political circles, Henry confronted with deliberate coolness. In Smithsonian publications, for example, he edited out all references to assessments and categorizations of culture based upon racial criteria. In part, Henry hoped not to alienate political representatives who controlled the Institution's appropriations. Individuals outside the Smithsonian pursued their own unique interpretations and examined the nature of race. Josiah Clark Nott and others sought evidence of the

inferiority of blacks and Indians. Such work, however, did not represent an institutionally based anthropological study, involving instead persons who fervently believed in the separation of races.

11. See also Nathan Reingold, ed., *The Papers of Joseph Henry*, vol. 1: *December 1797–October 1832: The Albany Years* (Washington, DC: Smithsonian Institution Press, 1972), 15–92; vol. 2: *November 1832–December 1835: The Princeton Years* (Washington, DC: Smithsonian Institution Press, 1975), 3, 261–66; vol. 3: *January 1836–December 1837: The Princeton Years*, (Washington, DC: Smithsonian Institution Press, 1979, 113–35; and vol. 4: *January 1838–December 1840: The Princeton Years*, (Washington, DC: Smithsonian Institution Press, 1981), 129–30.

12. The definition of ethnology changed over time. Historically oriented, nineteenth-century ethnology sought explanations for extant cultures, languages, and race. The twentieth–century variety has focused more on comparative studies of the past and the present. In the largest sense, ethnology is the scientific study of the origins and course of human culture. On the other hand, anthropology, broadly defined, encompasses the description, classification, history, and analysis of human culture. The two terms interchanged in the 1800s. See also Howard S. Miller, "Science and Private Agencies," in *Science and Society in the United States*, ed. D. Van Tassel and M. C. Hall (Homewood, IL: Dorsey Press, 1966), 195.

13. Hinsley, *Savages and Scientists*, 34–36, 40–42; Haskell, *Professional Social Science*, 67; Dupree, *Science in the Federal Government*, 81–87; Reingold, *Dictionary of Scientific Biography*, "Joseph Henry," 277–79; and Stanton, *Leopard's Spots*, 84–88.

14. Daniel G. Brinton, "American Languages and Why We Should Study Them," in Darnell, *History of Anthropology*, 206–17; Hinsley, *Savages and Scientists*, 42–67; Pliny Earle Goddard, "The Present Condition of Our Knowledge of North American Languages," in De Laguna, *Selected Papers from the American Anthropologist*, 385–402; Donald Collier and Harry Tschopik, Jr., "The Role of Museums in American Anthropology," *American Anthropologist* 56 (1954): 768–79; Dupree, *Science in the Federal Government*, 85–86; and Sally G. Kohlstedt, "A Step Toward Scientific Self-Identity in the United States: The Failure of the National Institute, 1844," in Reingold, *Science in America*, 79–103.

15. Kuhn, *Scientific Revolutions*, 17. See also Timothy H. H. Thoresen, "Art, Evolution, and History: A Case Study of Paradigm Change in Anthropology," *Journal of the History of the Behavioral Sciences* 13 (1977): 107–25. Dr. Thoresen makes several interesting points about the presence of theories of national and cultural evolution and their influence on the origins and evolution of Indian art. Beyond the technical and abstract uses of the various arts, he points out quite acutely that "local issues and institutions often shaped discussion of the topic in peculiarly national directions," (p. 109). A bit further on, he goes on to state, "American participation in the debate was constrained by a local cluster of issues which subordinated the art question to the more general problem of the origin and antiquity of man in America" (p. 109).

16. Donald Zochert, "Science and the Common Man in Ante-Bellum America," in Reingold, *Science in America*, 7–32; W. H. Holmes, "The World's Fair Congress of Anthropology," in De Laguna, *Selected Papers from the American Anthropologist*, 119–30; John B. Rae, "Science and Engineering in the History of America," *Technology and Culture* 2 (1961): 391–99; Garraty, *Commonwealth*, 95–96; and Kuhn, *Essential Tension*, 137–38.

17. Willey and Sabloff, *American Archaeology*, 86–92; George W. Stocking, Jr., *The Shaping of American Anthropology, 1883–1911: A Franz Boas Reader* (New York: Basic

Books, 1974), 1–18; Otis T. Mason, "The Technic of Aboriginal American Basketry," in De Laguna, *Selected Papers from American Anthropologist*, 559–78; and Wilcomb E. Washburn, *The Cosmos Club of Washington: A Centennial History, 1878–1978* (Washington, DC: Cosmos Club, 1978).

18. Hinsley, *Savages and Scientists*, 145–55; David J. Meltzer, "Archaeology in Transition: 19th Century Foundations and Early 20th Century Changes," paper presented at the Pacific Coast Branch, American Historical Association, Annual Meeting, San Diego, Calif., August 1983, 1–3. See also David J. Meltzer, "North American Archaeology and Archaeologists, 1879–1934," *American Antiquity* 50 (1985): 249–60. Meltzer details two important changes for archaeology: one, that a change in attitude resulted in the development of a class system within archaeology to control the amateur; and two, that the Bureau of American Ethnology was "the first agency of the federal government to make a substantial financial investment in American archaeology." The most important net result was a program that more closely defined an "archaeological science" (249–50, 258).

19. Wallace Stegner, "John Wesley Powell," in *Dictionary of Scientific Biography*, Vol. 11 (1970): 11820; Wallace Stegner, *Beyond the Hundredth Meridian: John Wesley Powell and the Second Opening of the West* (Boston: Houghton Mifflin, 1954), 1–30; Michael Ruse, *The Darwinian Revolution: Science Red in Tooth and Claw* (Chicago: University of Chicago Press, 1979), 36, 93, 152, 166, 249, 253, 263–64; "Obituary of John Wesley Powell," in *Selected Papers from American Anthropologist*, 136–37; George W. Stocking, Jr., "Some Problems in the Understanding of Nineteenth Century Cultural Evolutionism," in *Readings in the History of Anthropology*, 407; and Willey and Sabloff, *American Archaeology*, 79–81. Also of interest is Adam Kuper, "The Development of Lewis Henry Morgan's Evolutionism," *Journal of the History of the Behavioral Sciences* 21 (1985): 3–22.

20. Hinsley, *Savages and Scientists*, 151, 262–89; and Neil Judd, *The Bureau of American Ethnology: A Partial History* (Norman: University of Oklahoma Press, 1968).

21. Daniel Kevles observed, "Put mediocre scientists in control of a standards-setting institution, and the result will be mediocre standards." *The Physicists*, 38.

22. Meltzer, "Archaeology In Transition," 9–10.

23. Vesey, *Emergence of the American University*, 1–17, 23–119, 121–79. Some size of the discipline can be gauged by information derived from two sources. George W. Stocking, Jr., in his article, "Ideas and Institutions in American Anthropology: Thoughts Toward a History of the Interwar Years," contained in his edited work, *Selected Papers from the American Anthropologist: 1921–1945* (Washington, DC: American Anthropological Association, 1976), asserts that the size of the anthropological community by 1920 consisted of not more than three hundred members. Margaret W. Rossiter, in her book, *Women Scientists in America: Struggles and Strategies to 1940* (Baltimore: Johns Hopkins University Press, 1982), assembles data pertinent to the increasing number of women in science, also including the number within this diminutive realm. More women did graduate and the size of college enrollment went up as well. For example, in 1921, 108 male graduates in anthropology represented only 1.2 percent of 9,036 science graduates. By 1938, that number had increased to 225 out of 25,375, or 0.9 percent.

24. Vesey, *Emergence of the American University*, 175–79, 316–32, 342–59; Kevles, *The Physicists*, 7, 75–90; and Joseph Ben-David, "The Character and the Growth of Science in Germany and the United States," *Minerva* 7 (1968): 1–9.

25. Abram Kardiner and Edward Preble, *They Studied Man* (New York: Mentor Books, 1961), 117–20; "The Development of Anthropology," in De Laguna, *Selected Papers from the Anthropologist*, 102–103; George W. Stocking, Jr., "From Physics to Ethnology: Franz Boas' Arctic Expedition as a Problem in the Historiography of the Behavioral Sciences," *Journal of the History of the Behavioral Sciences* 1 (1965): 53–66; Fred W. Voget, "Franz Boas," in *Dictionary of Scientific Biography*, Vol. 2 (1970): 208; and Franz Boas, "Anthropology," in *Encyclopaedia of the Social Sciences*, 1 (1932): 73.

26. Ricardo Godoy, "Franz Boas and His Plans for an International School of American Archaeology and Ethnology in Mexico," *Journal of the History of the Behavioral Sciences* 13 (1977): 228–42; Voget, "Boas," 208–209; Kardiner and Preble, *They Studied Man*, 117–21, 132, 163–65; Alfred L. Kroeber, "A History of the Personality of Anthropology," in Darnell, *History of Anthropology*, 322–29; Alfred L. Kroeber "The Place of Anthropology in Universities," *American Anthropologist* 56 (1954);764–67; and John Freeman, "University Anthropology: Early Departments In The United States," *The Kroeber Anthropological Society Papers* 32 (1965): 78–90.

27. Willey and Sabloff, *American Archaeology*, 103–17; and Douglas R. Givens, *Alfred Vincent Kidder and the Development of Americanist Archaeology* (Albuquerque: University of New Mexico Press, 1992).

28. Willey and Sabloff, *American Archaeology*, 110–13; Gordon R. Willey, "One Hundred Years of American Archaeology," in Brew, *One Hundred Years of Anthropology*, 40–43; and Alfred V. Kidder, *An Introduction to the Study of Southwestern Archaeology* (New Haven: Yale University Press, 1924). Emil Haury originally brought this to my attention.

29. Willey and Sabloff, *American Archaeology*, 121–25. The work contains an excellent discussion on the influence of Kidder in this work and on Southwest archaeology. See also Richard B. Woodbury, *Alfred V. Kidder* (New York: Columbia University Press, 1973), and *Sixty Years of Southwestern Archaeology: A History of the Pecos Conference* (Albuquerque: University of New Mexico Press, 1993); Arthur C. Parker, *The Archaeological History of New York* (Albany: New York State Museum Bulletin, 1922); and Gerard Fowke, *Archaeological History of Ohio* (Columbus: Ohio State University Press, 1902).

30. John V. Murra "American Anthropology, the Early Years," in *American Anthropology: The Early Years*, ed. John v. Murra (New York: West Pub. Co., 1976), 1–8; and Regna Darnell, "Daniel Brinton and the Professionalization of American Anthropology," in Murra, *American Anthropology*, 69–98. The most prominent were and are—in Boston: the American Antiquarian Society, the Peabody Museum of Archaeology and Ethnology, the Boston Society of Natural History, and the Essex Institute; in New Haven: the American Oriental Society and the Connecticut Historical Society; in New York: the American Ethnological Society, the American Geographical Society, the American Museum of Natural History, and the Lyceum of Natural History; in Philadelphia: the American Philosophical Society, the Academy of Natural Sciences of Philadelphia, the Numismatic and Antiquarian Society, the Museum of Archaeology, and the Oriental Club of Philadelphia; and in Washington, DC: The Anthropological Society of Washington, the Washington Academy of Sciences, and the United States National Museum.

31. Robert E. Beider and Thomas B. Tax, "From Ethnologists to Anthropologists: A Brief History of the American Ethnological Society," in Murra, *American Anthropology*,

12–17; Alexander Lesser, "The American Ethnological Society: The Columbia Phase, 1906–1946," in Murra, *American Anthropology*, 126–35. See also, "American Museum of Natural History," in *The New Columbia Encyclopedia* (1975): 88; John R. Cole, "Nineteenth Century Fieldwork, Archaeology, and Museum Studies: Their Role in the Four-Field Definition of American Anthropology," in Murra, *American Anthropology*, 111–25; and Stocking, "Ideas and Institutions,", 1–2.

32. "Development of Anthropology," 93–103. George Stocking makes several points concerning the degree of coordination of the teaching of anthropology and archaeology. Many institutions offered instruction, but it varied considerably. Teaching obviously shifted from individual to individual. Where college instructors matriculated also proved critical, whether at Harvard or Columbia, for example. Harvard gave more specific training in archaeology than did the New York institution. Another factor depended on the departmental orientation of archaeology. A decided switch from museum-based archaeology occurred but once within the university, archaeology could affiliate with either social sciences or natural sciences. In 1931 at Yale anthropology was reorganized closely with sociology. Other universities witnessed an association with sciences such as geology, as at Louisiana State University. Anthropology even stood on its own and favored either one direction or the other according to the predilection of the members. Nonetheless, archaeology and anthropology had achieved a university base of operations. Stocking, "Ideas and Institutions," 11.

33. Cole, "Nineteenth Century Fieldwork," 120–24; Kardiner and Preble, *They Studied Man*, 122–24; and Stocking, "Ideas and Institutions," 2, 10. The debate over anthropology and patriotism ultimately resulted in Boas's censure by the AAA.

34. Collier and Tschopik, "The Role of Museums in American Archaeology," 768–79.

35. Neil Judd, "The Present Status of Archaeology in the United States," *American Anthropologist* 31 (1929): 401.

36. Ronald F. Lee, *United States: Historical and Archaeological Monuments* (Mexico, DF: Instituto Panamericano De Geographia E Historia, 1951), 19–21.

37. Judd, "The Present Status of Archaeology in the United States," 401–407; and Stocking, "Ideas and Institutions," 29–30. Stocking describes the growing concern with amateurism as a significant force that pushed archaeology toward greater control of research disciplinary coalescence. "Their concern with problems of professional self-identification was rather a defensive reaction against what they perceived as a massive threat from outside—what one of them since characterized as the 'rapidly spreading conflagration' of amateur archaeology. Relatively speaking, archaeology in this period suffered from an abundance of resources and a dearth of professional personnel."

Chapter 1

1. Glyn Daniel, "One Hundred Years of Old World Prehistory," in Brew, *One Hundred Years of Anthropology*, 59–60, 65–68, 73–74.

2. Jesse Jennings, et al., *The Native Americans: Ethnology and Backgrounds of the North American Indians*, 2nd ed.(New York: Harper & Row, 1977), ix–xix.

3. Yaron Ezrahi, "The Political Resources of American Science," *Science Studies* 1 (1971): 117–33. See also his later work, *The Descent of Icarus: Science and the Transforma-

tion of Contemporary Democracy (Cambridge: Harvard University Press, 1990), for a reciprocal effect.

4. Dupree, *Science in the Federal Government*, 294–300, 302–27.

5. Carl Guthe, "Twenty-Five Years of Archaeology in the Eastern United States," in *Archaeology of Eastern United States*, ed. James B. Griffin (Chicago: University of Chicago Press, 1952), 2.

6. Guthe, "Twenty-five Years of Archaeology," and interview with James B. Griffin, Ann Arbor, Michigan, August 1983.

7. Smithsonian Institution Archives, National Anthropological Archives, Records of the Bureau of American Ethnology, Miscellaneous Administrative Files, Box 3, "Scientific Minute Men in Anthropology" (hereafter referred to as SIA, NAA, BAE).

8. Guthe, "Twenty-Five Years," 4; and Richard B. Woodbury, "Regional Archaeological Conferences," *American Antiquity* 50 (1985): 434–44.

9. Guthe, "Twenty-five Years of Archaeology," 5, 9–10; interview with James B. Griffin; and W. C. McKern, "The Midwestern Taxonomic Methods As An Aid To Archaeological Culture Study," *American Antiquity* 4 (1939): 305.

10. McKern, "Midwestern Taxonomic Method," 306–13; and Albert C. Spaulding, "Fifty Years of Theory," *American Antiquity* 50 (1985): 301–308. Dr. Spaulding's excellent overview focuses on several key regional interpretations that were, in his words, "attempting to create a new archaeology, a scientific archaeology, in a context of an earlier prescientific, intuitive, and speculative archaeology." Spaulding, "Fifty Years of Theory," 302.

11. James B. Griffin, "Carl Eugen Guthe," *American Antiquity* 41 (1976): 168–77; and interview with James B. Griffin.

12. BAE, Additional Administrative Files, Box 4, Cooperative Ethnological Investigations File, letter McLeod to Abbot (7 April 1928).

13. BAE, Additional Administrative Files, Box 4, Cooperative Ethnological Investigations File, letter Guthe to Abbot (6 June 1928).

14. BAE, Additional Administrative Files, Box 4, Cooperative Ethnological Investigations File, letter Abbot to Guthe (6 June 1928).

15. BAE, Additional Administrative Files, Box 4, Cooperative Ethnological Investigations File, letter Abbot to Guthe (22 June 1928). The information and application criteria were scheduled to be published in a forthcoming issue of *Science*, among the most commonly read professional periodicals of the day. It enjoyed a much wider circulation than the *American Anthropologist*.

16. Interview with James B. Griffin; and interview with Gordon R. Willey, Harvard University, July 1982.

17. Lee, *Archaeological Monuments*, 19–27, 65–71.

18. BAE, Cooperative Ethnological Investigations File, letter Guthe to Stirling (n.d.). According to Stocking, "Ideas and Institutions," p. 12, the numbers of graduates for anthropology and archaeology fell by half in the 1930s. Ethnology and linguistics held their own in the Great Depression. The dominant percentage of male graduates appears grossly disproportionate in anthropology. However, Margaret Rossiter documents otherwise. Between 1920 and 1938, 59 women achieved the doctorate in anthropology out of a total of 197 granted, or 29.9 percent, the highest number in all science disciplines. Chemistry, for example, had only 8 percent female alumna. In addition, better than 80 percent of the female graduates found employment in colleges and universities, again among

the higher levels. Columbia University graduated the greatest number of female doctorates in anthropology. Its 9 degrees in 1938 alone constituted 45 percent of the total awards in the country.

On the other hand, the degree numbers for anthropology and archaeology graduates, as compared to the other sciences, was small. The basic reason was economic. Rossiter states: "The lack of jobs was a constant theme, almost a religion, in the 1920s and 1930s, as recruits were warned repeatedly to be ready for sacrifices." In this way, Franz Boas and his colleagues at Columbia University were able to transform the field's lack of jobs into a noble, self-sacrificing cause of great appeal to certain, especially affluent, men and women graduate students, as revealed in this interview with Dorothy Bramson Hammond, a Barnard graduate (class of 1939). When she entered the Columbia graduate school, she met with discouragement from all sides; it was, at that time, almost impossible to make a living in the field of anthropology. The graduate students were few in number, and quoting Mrs. Hammond: "To come in at all, they had to feel a real sense of dedication and a willingness to starve. I had the sense of dedication, all right, to say nothing of Papa, if it came to starvation. It is quite different now.... But then, in that small and very chummy Department, we all felt that we were doing something very special and exciting. And it was fun." Rossiter, *Women In Science*, 138–39, 152, 172–73, 181, 270.

19. Guthe, "Twenty-Five Years," 5–6; and see list of surveys in *American Anthropologist*, 24 (1922): 233–42; 25 (1923): 110–16; 27 (1925): 581–87; 28 (1926): 679–94; 29 (1927): 313–37; 30 (1928): 501–24; 31 (1929): 332–60; 32 (1930): 342–274; 33 (1931): 459–86; 34 (1932): 476–509; 35 (1933): 483–511; 36 (1934): 595–98. Regarding the hegemony of certain institutions, George Stocking states: "The role of the Columbia and Harvard departments was critical in the overall institutional life of the discipline. Between them there was a de facto division of labor, Harvard specializing in archaeology and physical anthropology, while Columbia took care of ethnology and linguistics. Together they produced thirty of the forty doctorates granted by 1920, with the others scattered among six different institutions, of which only the Universities of Pennsylvania and California (Berkeley) then maintained active instruction at the graduate level." Stocking, "Ideas and Institutions," 9.

20. Taylor, *A Study of Archaeology*, 46–94. See also James B. Griffin, "The Formation Of The Society For American Archaeology," *American Antiquity* 50 (1985): 261–71.

21. Fred Eggan, "Fay-Cooper Cole," *American Anthropologist* 65 (June 1963): 641–48; and Griffin, "Formation," 268. See also, George W. Stocking, Jr., *Anthropology at Chicago: Tradition, Discipline, Department* (Chicago: The Joseph Regenstein Library, 1980), 17–30.

22. See James T. Patterson, *The New Deal and the States: Federalism in Transition* (Princeton: Princeton University Press, 1969).

23. Guthe, "Twenty-Five Years," 6; BAE, Additional Administrative Files, letter Stirling to Steward (28 April 1933): "The Bureau appropriation for the coming fiscal year has been lopped off completely except in the sum to cover salaries, so that we are out as far as financing anything"; and SIA, Record Unit 46, Office of the Secretary, 1925–1949 (Charles D. Walcott, Charles G. Abbot, Alexander Wetmore) (hereafter referred to as SIA, Secretary), letter Van Hyning to Wetmore (11 December 1933).

24. Arthur M. Schlesinger, Jr., *The Age of Roosevelt*, vol. 1, *Crisis of the Old Order* (Boston: Houghton Mifflin Co., 1957), 1–32; William E. Leuchtenberg, *Franklin D. Roosevelt and the New Deal: 1932–1940* (New York: Harper & Row, Pubs., 1963), 1–17; William

E. Leuchtenberg, *The Perils of Prosperity: 1914–1932* (Chicago: University of Chicago Press, 1958), 241–73; and John K. Galbraith, *The Great Crash: 1929*, 3rd ed., (Boston: Houghton Mifflin Co., 1972).

25. C. Vann Woodward, *Origins of the New South: 1877–1913* (Baton Rouge: Louisiana State University Press, 1951), 396–436.

26. Interview with Gordon R. Willey; and interview with Marvin Kivett, Nebraska State Historical Society, Lincoln, Nebraska, February 1983.

27. Interview with Gordon R. Willey; National Archives, Record Group 69 (hereafter referred to as NA), FERA Files, File 100, "Genesis and History and Results of Federal Relief Programs," 1–3 ; and Southeastern Archaeological Archives, Society for Georgia Archaeology Files (hereafter referred to as SAA, SGA), uncataloged, Richard W. Smith, "A History of the Society for Georgia Archaeology," 14 April 1939 1–12, . The Fiftieth Anniversary issue of *American Antiquity* 50 (1985) contains two articles regarding amateurs in archaeology and societies with amateur membership. Both offer a substantial verification of the strength of public interest. See Carl H. Chapman, "The Amateur Archaeological Society: A Missouri Example," 241–48; Harold Mohrman, "Memoir Of An Avocational Archaeologist," 237–40; and David M. Gradwohl, "Marvin F. Kivett: 1917–1992," *American Antiquity* 59 (1994): 464–70.

28. Charles B. Hosmer, Jr., *Preservation Comes of Age: A History of the Preservation Movement before Williamsburg* (New York: Putnam Books, 1975); *Preservation Comes Of Age: From Williamsburg to the National Trust, 1926–1949* (Charlottesville: University of North Carolina Press, 1981); and interview with Gordon Willey.

29. NA, FERA files, File: FERA Grants to States, FERA State Files, Georgia, telegram Hopkins to Gay Sheppardson (15 December 1933); and Smith, "History of the Society," 13–15.

30. SIA, Secretary, letter Harris to Wetmore (14 February 1929), letter Harris to Wetmore (8 April 1929), letter Harrold to Wetmore (2 November 1932); and Smith, "History of the Society," 14–15.

31. Smith, "History of the Society," 15–16. Two forces affected the activities of the Society. These men were genuinely interested in "doing" good archaeology, but at the same time they exhibited a strong state orientation whereby they wished that Georgia relics remain in Georgia. In their minds, expeditions overseen by qualified personnel and a museum would assure both.

32. NAA, BAE, Administrative Files, letter Harrold to Stirling (27 July 1929), letter Stirling to Harrold (30 November 1932); SIA, Secretary, letter Harrold to Stirling (20 December 1932), letter Stirling to Harrold (26 January 1933), letter Harrold to Swanton (30 January 1933), letter Stirling to Harrold (7 November 1933). William Haag raises an important issue in his Fiftieth Anniversary article. In "Federal Aid To Archaeology In The Southeast, 1933–1942," *American Antiquity* 50 (1985): 272–79, Haag states the following regarding the museum: "Actually, their dream was to see the site become a tourist attraction and a source of municipal funds" (p. 277). Was the sole purpose for all this activity just for "crass commercialism?" The evidence suggests that the museum worked as a means to raise money from an interested public being educated at that museum. The funds would not only sustain the museum but underwrite further field research. Certainly, Dr. Haag may have formulated that feeling from his direct contact with these men. However, the written record does not support that view, nor do others interviewed for this study concur with Haag's statement. Perhaps Haag's belief reflects the long-

persistent conviction that any amateur involvement in archaeological work, much less the existence of nonacademically aligned museums, bode ill for "professional" archaeology. This criticism recurred in the 1930s and 1940s.

33. Interview with Gordon R. Willey; and interview with Marvin Kivett.
34. SIA, Secretary, letter Harris to Wetmore (14 February 1928). See Gordon R. Willey, *Portraits in American Archaeology: Remembrances of Some Distinguished Americanists* (Albuquerque: University of New Mexico Press, 1988), 243–66.
35. NAA, BAE, Administrative Files, letter Harrold to Stirling (27 July 1929), letter Stirling to Harrold (29 July 1929), letter Harrold to Stirling (28 November 1932); and SIA, Secretary, letter Stirling to Harrold (30 November 1932).
36. NA, WPA Files, Division of Information, Box 56, File 780-B, "A Study of American Prehistory under the Federal Works Program," 2–3; interview with Carlyle S. Smith, Lawrence, Kansas, February 1983; and SIA, Secretary, letter Harrold to Swanton (30 January 1933).
37. "American Prehistory," 2–3; and interview with Carlyle S. Smith.
38. NAA, BAE, Administrative Files, letter Harrold to Stirling (28 November 1932).
39. NAA, BAE, Administrative Files, letter Stirling to Harrold. (30 November 1932); and SIA, Secretary, letter Harrold to Stirling (20 December 1932), letter Stirling to Harold (26 January 1933).
40. NAA, BAE, Administrative Files, letter Harrold to Swanton (30 January 1933).
41. SIA, Secretary, letter Stirling to Harrold (7 February 1933).
42. Leuchtenberg, *New Deal*, 41–62; and Schlesinger, *The Age of Roosevelt*, vol. 2, *The Coming of the New Deal* (Boston: Houghton Mifflin Co., 1959), 1–34.
43. SIA, Secretary, letter Harrold to Stirling (10 October 1934).
44. Lee, *Archaeological Monuments*, 11–18, 68–70.
45. SIA, Secretary, letter Stirling to Harrold (20 December 1934).
46. Ibid.
47. Guthe, "Twenty-Five Years," 7–8.

Chapter 2

1. George I. Quimby, Jr., "A Brief History of WPA Archaeology," in *The Uses of Anthropology*, ed. Walter Goldschmidt (Washington, DC: American Anthropological Association, 1979), 110.
2. Leuchtenberg, *New Deal*, 41–62, 121–24; and William W. Bremer, "Along the 'American Way': The New Deal's Work Relief Programs for the Unemployed," *Journal of American History* 62 (1975): 636–52.
3. Leuchtenberg, *New Deal*, 71–72; Schlesinger, *Coming of the New Deal*, 425–33; and Gerald D. Nash,"Herbert Hoover and the Origins of the Reconstruction Finance Corporation," *Mississippi Valley Historical Review* 46 (1959): 455–68.
4. Leuchtenberg, *New Deal*, 123–25; Schlesinger, *Coming of the New Deal*, 265–94; and Florence Peterson, "CWA: A Candid Appraisal," *Atlantic Monthly* 153 (1934): 587–90.
5. Robert Sherwood, *Roosevelt and Hopkins: An Intimate History*, rev. ed. (New York: Harper & Row, Pubs., 1948), 3–49; Paul A. Kurzman, *Harry Hopkins and the New Deal*

(Fair Lawn, NJ: R. E. Burdick, Inc., 1974), 1–12; Henry H. Adams, *Harry Hopkins* (New York: G. P. Putnam's Sons, 1977), 1–132; and George McJimsey, *Harry Hopkins: Ally of the Poor and Defender of Democracy* (Cambridge: Harvard University Press, 1987), 1–83. The last biography stands as the most recent and scholarly treatment of Hopkins's life.

6. *National Industrial Recovery Act, Statutes at Large*, sec. 67, 200–202 (1934); Harry Hopkins Papers, CWA, "The Federal Civil Works Administration: A study covering its organization in November 1933 and its operations until 31 March 1934," 1–2; and Leuchtenberg, *New Deal*, 70–71.

7. "The Federal Civil Works Administration," 2–3.

8. Public Works Administration, *A Four Year Record of the Construction of Permanent and Useful Public Works* (Washington, DC: Government Printing Office, 1937); and National Resources Planning Board, *The Economic Effects of the Federal Public Works Expenditures: 1933–1938* (Washington, DC: Government Printing Office, 1940).

9. Sherwood, *Roosevelt and Hopkins*, 3–23; and McJimsey, *Hopkins*, 44–61. For a fuller discussion of the administration as a whole, see William D. Reeves, "PWA and Competitive Administration in the New Deal," *Journal of American History* 60 (1973): 357–72.

10. Schlesinger, *Coming of the New Deal*, 270–78; John A. Garraty, *The Great Depression: An Inquiry into the causes, course, and consequences of the Worldwide Depression of the Nineteen-Thirties as seen by contemporaries and in the light of History* (New York: Harcourt Brace Jovanovich Pubs., 1986), 151, 158–59; and Albert U. Romasco, *The Politics of Recovery: Roosevelt's New Deal* (New York: Oxford University Press, 1983). Also see James T. Patterson, *Congressional Conservatism and the New Deal: The Growth of the Conservative Coalition in Congress, 1933–1939* (Lexington: University of Kentucky Press, 1967). Patterson discusses how Roosevelt identified the members of the white hierarchy in the South. He then attempted to work around them by expanding his broad-based constituency through a variety of programs (pp. 1–127). No documentation was found that ties archaeology with this political effort. Considering the small part played by these programs, it is doubtful any exists.

11. Sclesinger, *Coming of the New Deal*, 270–78.

12. National Planning Board, *Final Report: 1933–34* (Washington, DC: Government Printing Office). Later criticisms by archaeologists regarding relief archaeology often centered on the primacy of relief considerations rather than the "scientific" criteria used in the selection process. The circumstances were complex. During the CWA the Smithsonian chose from literally thousands of project applications around the nation. Those culled conformed to the relief strictures, but in any case were picked based upon their own scientific validity. It was a matter of reconciling two imperatives. Later relief operations usually met these standards as well. Some of the criticism may have originated in continued hostility toward the Smithsonian and an incomplete understanding of how officials derived and operated actual relief policies. Other critics directed similar derision at the make-work "leaf-raking" jobs associated with the government.

13. Leuchtenberg, *New Deal*, 33, 120–27.

14. "The Federal Civil Works Administration," 2–8.

15. SIA, Secretary, letter Stirling to Harrold (30 November 1932).

16. NAA, WPA Monthly—Quarterly Reports: Louisiana (hereafter referred to as WPA Monthly Reports), "Louisiana State University—Work Projects Administration No. 165-64-59."

17. Ibid., 1.

18. Ibid. For a further explanation and overview, see John W. Bennett, *Archaeological Explorations in Jo Daviess County, Illinois: The Work of William Baker Nickerson (1895–1901) and the University of Chicago (1926–32)* (Chicago: University of Chicago Press, 1945); Fay-Cooper Cole and Thorne Deuel, *Rediscovering Illinois: Archaeological Explorations In And Around Fulton County* (Chicago: University of Chicago Press, 1937); and Spaulding, "Fifty Years," 302. Jesse Jennings illuminated me on these two works.

19. Waldo R. Wedel, "Archaeological Investigations At Buena Vista Lake, Kern County, California," *Bureau of American Ethnology Bulletin* 130 (1941): 1.

20. Frank M. Setzler, "Archaeological Explorations in the United States: 1930–1942," *Acta Americana* 1 (1943): 206–208; and NAA, James A. Ford Papers, uncataloged, Box 3, Marksville. See also Gordon R. Willey's wonderful account of James Ford in *Portraits in American Archaeology*, 51–74.

21. Frank M. Setzler and W. Duncan Strong, "Archaeology and Relief," *American Antiquity* 1 (1936): 301–309.

22. SIA, Secretary, letter Harrold to Wetmore (7 December 1933); letter Wetmore to Ford (15 December 1933); memorandum Graf to Stirling (24 November 1933); press release (26 December 1933); and NAA, Frank M. Setzler Papers (hereafter referred to as FMS), Series 1, subseries 2, Box 11, WPA Correspondence, File 8, "Brief Summary of CWA Archaeological Projects."

23. Matthew W. Stirling, "Smithsonian Archaeological Projects Conducted under the Federal Emergency Relief Administration, 1933–34," *Annual Report of the Board of Regents* (Washington, DC: Smithsonian Institution, 1934), 371–401.

24. SIA, Secretary, letter Stirling to Harrold (8 December 1933); and letter Harrold to Stirling (9 December 1933).

25. SIA Secretary, letter Stirling to Harrold (7 December 1933).

26. Ibid.

27. SIA Secretary, letter Harrold to Swanton (3 December 1933).

28. SIA Secretary, letter Harrold to Stirling (9 December 1933).

29. Stirling, "Smithsonian Archaeological Projects," 371; and Setzler and Strong, "Archaeology and Relief," 301.

30. Stirling, "Smithsonian Archaeological Projects," 371–72. Setzler dominated in the daily representation of archaeology and anthropology. Secretary Wetmore testified on behalf of archaeology and the remainder of the Institution before Congress.

31. Ibid., 398–99; "Summary of CWA Projects," 3; and "American Prehistory," 20–21. See also Gordon R. Willey, *Portraits in American Archaeology*, 75–98, for another insightful essay on a major figure in American archaeology.

32. Stirling, "Smithsonian Archaeological Projects," 398–99; "American Prehistory," 20–21; and "Summary of CWA Projects," 3.

33. NAA, FMS, Series 1, Subseries 2, U.S. National Museum (USNM), Department of Anthropology, Miscellaneous Administrative Records, 1896–1965, CWA File, Box 1, series 1–2, "Tulamniu CWA Project," 6–7.

34. Leuchtenberg, *New Deal*, 130–33; Leuchtenberg, *Perils of Prosperity*, 250–73; and Schlesinger, *Coming of the New Deal*, 308–15.

35. "Tulamniu CWA Project," 7.

36. Ibid.

37. Ibid., p. 22.
38. Wedel, "Archaeological Investigations at Buena Vista Lake," 156.
39. Ibid., 1–17.
40. Stirling, "Smithsonian Archaeological Projects," 392–94; "American Prehistory," 15–17; "Tulamniu CWA Project," 7–8; and Frank M. Setzler and Jesse D. Jennings, "Peachtree Mound and Village Site, Cherokee County North Carolina," *Bureau of American Ethnology Bulletin* 131 (1941): 1–5.
41. Setzler and Jennings, "Peachtree Mound and Village Site," ix, 13, 52–57; and Jesse D. Jennings, *Accidental Archaeologist: Memoirs of Jesse D. Jennings*, forward by C. Melvin Aikens (Salt Lake City: University of Utah Press, 1994), 73–77.
42. Michael D. Coe, "Matthew W. Stirling, " *American Antiquity* 41 (1976): 37–43.
43. FMS, USNM, CWA File, "Florida CWA Project," 1.
44. FMS, USNM, File 128726–128752, letter Hall to Roosevelt (25 January 1934), letter Wetmore to Hall (3 March 1934); and SIA, Secretary, letter Van Hyning to Wetmore (11 December 1933). Edwin Lyon asserted in a 1982 dissertation that Stirling remained in Washington to direct the digs. The following group of letters to and from Stirling disprove this. See Edwin Austin Lyon II, "New Deal Archaeology in the Southeast: WPA, TVA, NPS, 1934–1942" (Ph.D. dissertation, Louisiana State University, 1982), 39.
45. SIA, Secretary, letter Wetmore to C. B. Treadway (12 December 1933).
46. SIA, Secretary, letter Wetmore to Stirling (13 December 1933).
47. SIA, Secretary, letter Wetmore to Van Hyning (13 December 1933).
48. SIA, Secretary, letter Van Hyning to Wetmore (15 December 1933).
49. FMS, USNM, File 128726–128752, letter Stirling to Setzler (17 December 1933).
50. Schlesinger, *Coming of the New Deal*, 273–77.
51. FMS, USNM, File 128726–128752, letter Stirling to Setzler (17 December 1933).
52. Ibid.
53. FMS, USNM, File 128726-1288752, and letter Stirling to Setzler (20 December 1933).
54. FMS, USNM, File 128726-1288752, letter Setzler to Stirling (27 December 1933).
55. FMS, USNM, File 128726-1288752, letter Stirling to Van Hyning (28 December 1933).
56. FMS, USNM, File 128726-1288752, letter Stirling to Anthony (29 December 1933).
57. FMS, USNM, File 128726-1288752, letter Setzler to Stirling (25 January 1934), 1.
58. FMS, USNM, File 128726-1288752, 2.
59. FMS, USNM, File 128726-1288752, letter Stirling to Wetmore (30 January 1934), 1–2.
60. Ibid.
61. Ibid.; and interview with Gordon R. Willey.
62. Stirling, "Smithsonian Archaeological Projects," 372–76; "American Prehistory," 4–6; and "Florida CWA Project," 1–2.
63. Stirling, "Smithsonian Archaeological Projects," 372–76; and "American Prehistory," 4–6.
64. Stirling, "Smithsonian Archaeological Projects," 372–76; and "American Prehistory," 4–6.

65. "American Prehistory," 6; Stirling, "Smithsonian Archaeological Projects," 383–85; and "Florida CWA Project," 3–4.
66. "American Prehistory," 6; Stirling, "Smithsonian Archaeological projects," 383–85; and "Florida CWA Project," 3–4.
67. "American Prehistory," 9; and Stirling, "Smithsonian Archaeological Projects," 384–85.
68. "American Prehistory," 6–9; and Stirling, "Smithsonian Archaeological Projects," 378–83.
69. "Florida CWA Projects," 7–9; Stirling, "Smithsonian Archaeological Projects," 385–88; "American Prehistory," 9–11; and "Summary of CWA Projects," 1.
70. Stirling, "Smithsonian Archaeological Projects," 388–89; and "American Prehistory," 11–13; and Jennings, *Accidental Archaeologist*, 78–83.
71. Gordon R. Willey, "Archaeology on the Florida Gulf Coast," *Smithsonian Miscellaneous Collections* 113 (1949).
72. Stirling, "Smithsonian Archaeological Projects," 391–92.
73. FMS, USNM, CWA File, "Georgia CWA Project," 1. See also Gordon R. Willey, *Portraits in American Archaeology*, 27–50.
74. Interview with James B. Griffin.
75. Ibid.; and Willey and Sabloff, *American Archaeology*, 107, 114–16.
76. Willey and Sabloff, *American Archaeology*, 119–20.
77. Ibid., 120–21; and Spaulding, "Fifty Years," 305–306.
78. "American Prehistory," 13–14; Stirling, "Smithsonian Archaeological Projects," 389–90; and "Georgia CWA Project," 2.
79. Stirling, "Smithsonian Archaeological Projects," 391.
80. Ibid., 390–92; and "American Prehistory," 14–15.
81. SIA, Secretary, letter Harrold to Stirling (9 December 1933).
82. Ibid.
83. Ibid.
84. Ibid.
85. NA, FERA, State Files, Georgia, "A History of the CWA Administration, 1933–34," ed. Gilbert H. Boggs, 48–49; and interviews with Gordon R. Willey and James B. Griffin.
86. SIA, Secretary, letter Lane to Hopkins (5 February 1934); FMS, WPA Correspondence, Box 10, File 1, letter Kelly to Chatelain (5 August 1935), letter Harrold to Stirling (10 October 1934), letter Stirling to Harrold (12 October 1934); Box 2, General Correspondence, "An Open Letter to the Georgia Delegation in Congress from the Society for Georgia Archaeology"; and NA, FERA, State Files, Georgia, letter Vinson to Hopkins (5 June 1934), letter Kelly to Wetmore (2 January 1934).
87. "American Prehistory," 17.
88. FMS, USNM, Department of Anthropology, Official Files 1899–1959, Box 7, letter Cox to Abbot (27 January 1932).
89. "American Prehistory," 17–19; Stirling, "Smithsonian Archaeological Projects," 394–98; FMS, USNM, CWA file, "Tennessee CWA Project," 1–3; and "Summary of CWA Projects," 3.
90. Stirling, "Smithsonian Archaeological Projects," 397.
91. FMS, USNM, Department of Anthropology, CWA File, letter Setzler to Stirling (25 January 1934).

92. Leuchtenberg, *New Deal*, 12, 54–55; Schlesinger, *Crisis of the Old Order*, 538; and Schlesinger, *Coming of the New Deal*, 320–40.
93. Leuchtenberg, *New Deal*, 165; and Schlesinger, *Coming of the New Deal*, 326.
94. SIA, Secretary, letter Stirling to Guthe (10 August 1933).
95. SIA Secretary, letter Poffenberger to Draper (30 August 1933); and "Report of the Science Advisory Board, July 31, 1933."
96. SIA, Secretary, letter Stirling to Guthe (10 August 1933). The Science Service, a nonprofit foundation set up in Washington, D.C., to popularize science, represented various academic scientific societies.
97. NAA, BAE, Administrative Files, TVA, Boxes 33–35; William S. Webb, "An Archaeological Survey of Wheeler Basin on the Tennessee River in Northern Alabama, " *Bureau of American Ethnology Bulletin* 122 (1939): 2–3; and "An Archaeological Survey of the Norris Basin in Eastern Tennessee," *Bureau of American Ethnology Bulletin* 118 (1938): 2–3.
98. Douglas W. Schwartz, *Conceptions of Kentucky Prehistory: A Case Study in the History of Archaeology* (Lexington: University of Kentucky Press, 1967), 31–54; William G. Haag, "William Snyder Webb, 1882–1964" *American Antiquity* 30 (1965): 470–73; and NAA, BAE, Cooperative Ethnological Investigations File, letter Webb to Abbot (21 January 1930), memorandum Abbot to Hough (26 January 1930), memorandum Dorsey to Bryant (31 January 1930), and letter Abbot to Webb (4 February 1932).
99. Schwatz, *Conceptions*, 31–54; and Haag, "Webb," 470–73.
100. Carl Guthe, "Twenty-Five Years," 6; Webb, "Wheeler Basin," 2; and "Norris Basin," 7. An important point about the institutional state of archaeology merits discussion. Academic archaeologists had achieved a newfound success with the inauguration of the CWA and the TVA projects. Still, the discipline was not devoid of the influence of talented amateurs. The dearth of trained personnel remained in force throughout the 1930s. Proficient, non-university trained archaeologists, such as William Webb, Walter Jones, and T. M. N. Lewis, due to their reputations and accomplishments, were placed in charge of large salvage projects. Such circumstances further attested to the true emergency nature of New Deal archaeology and the slow growth of a fully professionalized discipline.
101. Webb, "Norris Basin," 362–64; and FMS, WPA Correspondence, File 7, letter Willey to Setzler (3 April 1934).
102. Michael S. Holmes, *New Deal in Georgia: An Administrative History* (Westport, CT: Greenwood Press, 1975), 61–95; and Warren C. Whatley, "Labor for the Picking: The New Deal in the South," *Journal of Economic History* 43 (1983): 905–29.
103. Setzler and Strong, "Archaeology and Relief."
104. Ibid.
105. "An Open Letter to the Georgia Delegation;" and Boggs, "History of the CWA Administration," 9–10.
106. FMS, CWA File, File 5, "Georgia CWA Project," (9 February 1934).
107. FMS, CWA File, File 5, letter Kelly to Wetmore (24 December 1933).
108. FMS, CWA File, File 6, letter Ford to Setzler (28 December 1933).
109. SIA, Secretary, letter Abbot to Stone (26 January 1934).
110. Frank M. Setzler, Introduction to "Archaeology of the Funeral Mound Ocmulgee National Monument, Georgia," *Archaeological Research Series No. 3, National Park Service* 3 (1956): 1.

Chapter 3

1. Schlesinger, *Coming of the New Deal*, 268–70. Hopkins did not care for sustained direct relief. Work relief, he believed, "preserves a man's morale." Roosevelt echoed a like concern. See also William W. Bremer, "Along the American Way: The New Deal's Work Relief Programs for the Unemployed," *Journal of American History* 62 (1975): 636–37.

2. Leuchtenberg, *New Deal*, 123–25; and Schlesinger, *Politics of Upheaval*, 389–92; Willey and Sabloff, *American Archaeology*, 44–45; and Voget, "Boas," 209. The official bureaucratic account of the FERA digs, "American Prehistory," faintly negative in tone, maintained that the digs failed to achieve the highest academic standard. The anonymous author offered no thorough explanation for this assessment. A likely reason stemmed from the fact that the Smithsonian had been largely excluded from project proposal and control after CWA. The study did end on a positive vein and concluded that the archaeological information garnered was important (pp. 2–3, 22)

3. The academic archaeological community failed to perceive this dilution of Smithsonian oversight and lack of centralized direction until almost 1937. No evidence has been uncovered that contradicts these views. Academic archaeological leaders proved more insightful in the post-Depression era.

4. Carl Guthe, "Reflections On The Founding Of The Society For American Archaeology," *American Antiquity* 32 (1967): 436; and Stocking, "Ideas and Institutions," 1–4, 10–12.

5. Guthe, "Reflections," 437–38; "The Society for American Archaeology Organizational Meeting," *American Antiquity* 1 (1935): 141; and James B. Griffin, "The Formation Of The Society For American Archaeology," *American Antiquity* 50 (1985): 261–63.

6. "Organizational Meeting," 142–43.

7. "Organizational Meeting," 142–43.

8. "Organizational Meeting," 141–51; Griffin, "Formation," 436–38; and Stocking, "Ideas and Institutions," 30.

9. "Organizational Meeting," 141–51. Fay-Cooper Cole voiced the concern of the parent anthropological community about a separatist movement. He thought such an action acceptable if it simply constituted a search for a stronger voice in the determination and solution of specific needs as well as the creation of an independent journal. Other elements of the anthropological discipline, such as linguistics and physical anthropology, possessed publication outlets of their own in addition to representation in the main anthropological journal, the *American Anthropologist*. Archaeologists expressed their determination to remain within the disciplinary purview of anthropology but wanted their own journal to address the increasing amount of information amassed.

10. President—A. C. Parker, Rochester Museum of Arts and Sciences, New York; Vice President—M. R. Harrington, Southwest Museum, Los Angeles; Editor—W. C. McKern, Milwaukee Public Museum; Secretary/Treasurer—Carl Guthe, Chairman of the Committee on State Archaeological Surveys; Council Members—E. F. Greenman, E. W. Haury, D. Jenness, F. H. H. Roberts, Jr., Leslie Spier, W. D. Strong, G. C. Valliant, and W. S. Webb.

11. Webb, "Norris Basin," 2–3. Florence Hawley divided her time between these

two campuses for her tree-ring research. Additionally, the University of Arizona also provided laboratory facilities to corroborate research findings.

12. Willey and Sabloff, *American Archaeology*, 1–6, 125–26; and Freeman, "University Anthropology."

13. Webb, "Norris Basin," 263, 363–82.

14. Andrew H. Whiteford, "A Frame Of Reference For The Archaeology Of Eastern Tennessee," in *Archaeology of the Eastern United States*, ed James B. Griffin (Chicago: University of Chicago Press, 1952), 207–25.

15. W. D. Funkhouser, "A Study Of The Physical Anthropology And Pathology Of The Osteological Material From The Norris Basin," in Webb, "Norris Basin," 248.

16. James B. Griffin, "The Ceramic Remains From Norris Basin, Tennessee," in Webb, "Norris Basin," 253–309.

17. Florence M. Hawley, "Tree Ring Dating For Southeastern Mounds," in Webb, "Norris Basin," 2, 357–59, 362.

18. Webb, "Wheeler Basin," 1–4, 9–20. Walter Jones, active in amateur archaeology in the 1920s, sustained a close relationship with the Society for Georgia Archaeology. He conducted archaeologic work for the Alabama Museum throughout this period. Later, trained specialists assumed his duties.

19. Guthe, "Reflections," 437.

20. Webb, "Wheeler Basin," 5–7.

21. Ibid., 201.

22. See James B. Griffin, "Prehistoric Cultures Of The Central Mississippi Valley," in Griffin, *Eastern Archaeology*, 226–38.

23. W. D. Funkhouser, "A Study Of The Physical Anthropology And Pathology Of The Osteological Material From The Wheeler Basin," in Webb, "Wheeler Basin," 109–26.

24. James B. Griffin, "Report On The Ceramics Of Wheeler Basin," in Webb, "Wheeler Basin," 127–65.

25. Setzler and Strong, "Archaeology and Relief," 306–309.

CHAPTER 4

1. Leuchtenberg, *New Deal*, 124–25; and Schlesinger, *Coming of the New Deal*, 2–94.

2. Arthur W. MacMahon, John D. Millet, and Gladys Ogden, *The Administration of Federal Work Relief* (New York: Columbia University Press, 1946; reprint ed., New York: Da Capo Press, 1971), 1–27, 44–70; and Leuchtenberg, *New Deal*, 125.

3. MacMahon, *Work Relief*, 70–78; and Leuchtenberg, *New Deal*, 91, 124–30.

4. MacMahon, *Work Relief*, 153–55.

5. Leuchtenberg, *New Deal*, 124–25.

6. MacMahon, *Work Relief*, 66–86.

7. Schlesinger, *Coming of the New Deal*, 271–75.

8. Regarding the role of the Smithsonian during the WPA, a major interpretive difference exists between Edwin Lyon and this author. Lyon contends in his dissertation, on pp. 69–70, that only the Institution "approved" archaeological projects. In addition, he maintains that Frank Setzler "may" have seen every proposal and, therefore, exercised considerable influence on the direction of relief archaeology. The single document cited

as the source for these statements is a letter from William Webb to James DeJarnette (9 February 1938). One sentence appears to be crucial to his interpretation: "All archaeological projects of WPA must be approved by him [Setzler] individually." Several points deserve discussion here. First, this specific evidence may or may not indicate the exact administrative role filled by the Smithsonian. Webb, a field supervisor not necessarily privy to accurate information about internal WPA administrative policy, apparently believed that the Smithsonian and, in particular, Setzler sanctioned each and every application. In truth, the Institution did review many. However, the specific, allotted power to approve or disapprove lay only within the Works Progress Administration itself. Second, evidence presented later in this chapter (for example, a key letter from Director Demeray of the NPS to Harry Hopkins) indicates not only that the NPS also had a reviewing status, but that state WPA administrators often failed to follow proper procedure directing them to consult and check with either government agency concerning archaeologic projects. All information viewed together reveals that considerable confusion and misunderstanding of procedure and appropriate decision-making channels existed. This was the very point that Demeray wished to make. A major contention of this study is that such disorder led to a restructuring of WPA review and application policy; and those changes further impacted the expanding influence of academic archaeologists (both government and college).

Finally, since Webb and other archaeologists believed that the Smithsonian was singly responsible for review and approval of archaeological relief projects, it may explain to some degree the hostility directed toward the Institution after 1939. Members of the Committee on Basic Needs and the Planning Committee felt that the Smithsonian was a factor in the overall lack of high quality in relief archaeologic programs. Subsequent actions taken by the aforementioned Committees (and the later Committee for the Recovery of Archaeological Remains) to isolate and control the Smithsonian may have been based partially on such assumptions. Strong personal animosities against Setzler and Secretary Wetmore were also a factor.

9. MacMahon, *Work Relief*, 153–55.

10. Ibid., 89–90.

11. Ibid., 98–99, 129.

12. Ibid., 75.

13. NA, WPA, Women's and Professional Division, Box 57, File 780-B, 1938, "Monthly Reports, September, 1938," 17–22.

14. Interview with Carlyle S. Smith, Lawrence, Kansas, February 1983.

15. William S. Webb Papers (hereafter referred to as Webb Papers), University of Kentucky, Box 3, "WPA," letter Fullerton to Webb (28 March 1938).

16. NA, WPA, *Operating Procedure No. 4* (Washington, DC: Government Printing Office, 1936).

17. FMS, WPA Correspondence, Box 13, letter Demeray to Hopkins (2 October 1937).

18. NA, WPA, *Operating Procedure No. 4* (Washington, DC: Government Printing Office, 1938). See, especially, sections 1–5.

19. Ibid.

20. Letter Petrullo to author (3 August 1987), 1–2; and "Vincenzo Petrullo," in *American Men and Women of Science* (1973): 1929. Rice formally worked as a professor of social science even though he trained as an anthropologist. It was in this capacity that he knew of Petrullo's capabilities. Also recall that all archaeologists were trained as anthro-

pologists. Archaeology stood as a specialty, like ethnology. General cultural presumptions and methodology overlapped into all areas.

21. Webb Papers, "WPA," letter Petrullo to Webb (22 April 1938).

22. NA, WPA, Division of Information, File 780-B, "General Notes on WPA Archaeological Projects obtained in interview 2/9/39 with Dr. Vincenzo Petrullo, national consultant," 9 February 1939; and "American Prehistory," 30–31.

23. Webb Papers, "WPA," letter Woodward to Goodman (21 March 1938).

24. Webb Papers, "WPA," letter Woodward to Goodman (12 June 1937). A similar letter went out about a year earlier to Goodman from the Director of the Women's and Professional Projects, Elizabeth Fullerton. Woodward requested supplementary data about a project in order to review an application. She also solicited more detailed information about the work: planned excavation techniques, an account of the archaeological questions to be addressed and their relation to other problems, and specification about both the provisions for storage of specimens and publication of research results. The letter also alerted the applicants about what their related responsibilities would be after project approval occurred. Written consent would be obtained from the National Park Service after fulfillment of the other stipulations. Representative documentation, including photographs and written descriptions, would be forwarded to the Smithsonian. The sponsor would submit to the WPA periodic reports with appropriate photos and sketches. A complete report would be forthcoming upon completion of the project. In addition, a specific plan for publication would accompany it.

25. "Petrullo Interview"; and "American Prehistory."

26. Verne Chaetlain, "The National Park Service and the New Deal," *OAH Newsletter* 13 (February 1985): 11–13. For a varied recount of the history of the Park Service's involvement in the preservation movement, one which ignores the importance of the 1933 Executive Order, see A. R. Kelly, "Archaeology In The National Park Service," *American Antiquity* 4 (1940): 274–82.

27. Chaetlain, "The National Park Service," 11–13; and Kelly, "Archaeology in the National Park Service," 274–82.

28. Chaetlain, "The National Park Service," 11–13; and Kelly, "Archaeology in the National Park Service," 274–82.

29. FMS, WPA Correspondence, Box 11, letter Kelly to Setzler (4 January 1937), letter Kelly to Setzler (10 June 1937).

30. FMS, WPA Correspondence, Box 10, memorandum Stirling to Setzler (15 June 1940).

31. Webb Papers, "WPA," letter Russell to Lee (23 March 1938).

32. SIA, Secretary, letter Stirling to Woodward (18 June 1937), and letter Woodward to Stirling (1 July 1937).

33. SIA, Secretary, memorandum Setzler to Akers (26 October 1937).

34. SIA, Secretary, letter Abbot to Ickes (5 April 1938), letter Interior to Abbot (4 May 1938).

35. SIA, Secretary, letter Ickes to Abbot (10 March 1939), letter Abbot to Ickes (18 March 1939).

36. For some examples, see NAA, Frederick Johnson Papers, File: "Planning Committee, SAA Reports, Minutes and Other Documents," "Planning Committee, Notes on Meeting of January 8–13, 1945," 2.

CHAPTER 5

1. WPA, Division of Information, box 56, File 780-A.
2. Guthe, "Twenty-Five Years," 6–10; "Society for American Archaeology," 439–40; and Frederick Johnson, "A Quarter Century of Growth in American Archaeology," *American Antiquity* 27 (1961): 1–6.
3. Interview with James B. Griffin; Guthe, "Twenty-Five Years," 6–10; National Research Council, "Conference on Southern Pre-history Held under the Auspices of the Division of Anthropology and Psychology, Committee on State Archaeological Surveys, National Research Council, Birmingham, Alabama, December 18, 19, and 20, 1932," Washington; and "The Indianapolis Archaeological Conference Held under the Auspices of the Division of Anthropology and Psychology, Committee on State Archaeological Surveys, National Research Council, Indianapolis, Indiana, December 6, 7, and 8, 1935," Washington. An article by Richard B. Woodbury comes to similar conclusions; see "Regional Archaeological Conferences," *American Antiquity* 50 (1985): 434–44.
4. William G. Haag, "Twenty Five Years of Eastern Archaeology," *American Antiquity* 27 (1961): 16.
5. Gordon R. Willey, "James Alfred Ford, 1911–1968," *American Antiquity* 24 (1969): 63–64; and Clifford Evans, "James Alfred Ford, 1911–1968," *American Anthropologist* 70 (1968): 1162. See also Gordon Willey, *Portraits in American Archaeology*, 51–74.
6. W. D. Strong, "Recommendations of the Committee on Basic Needs in American Archaeology," *Science* 90 (December 1939): 528–30. Although archaeology might appear to have been randomly "dumped" into a "catch-all" bureaucratic division, the outcome was productive. The women administrators in charge were all college-trained, sympathetic, and supportive of the academic endeavors of archaeologists. Archaeology might not have fared so well in a more systematic, work-oriented building section.
7. Leuchtenberg, *New Deal*, 53, 57–58; and Schlesinger, *Coming of the New Deal*, 184–90.
8. Robert W. Neumann, *An Introduction to Louisiana Archaeology* (Baton Rouge: Louisiana State University Press, 1984), 45.
9. NA, WPA, Division of Information, File 780-A, "The LSU–WPA Archaeological Survey of Louisiana"; "WPA Index"; WPA Monthly Reports: Louisiana, #30,064, #50,045.
10. Interview with Gordon R. Willey; interview with Carlyle S. Smith; letter George I. Quimby, Jr., to author (September 1983); and WPA, "LSU–WPA." See also Gordon R. Willey, *Portraits in American Archaeology*, 3–26.
11. Neumann, *Louisiana Archaeology*, 302–305.
12. Ibid., 48–51; Winslow W. Walker, "A Reconnaissance of Northern Louisiana Mounds," *Explorations and Field-Work of the Smithsonian Institution in 1931* (Washington, DC: Government Printing Office, 1931): 169–74; and Winslow W. Walker, "Trailing the Moundbuilders of the Mississippi Valley," *Explorations and Field-Work of the Smithsonian Institution in 1932* (Washington, DC: Government Printing Office, 1932): 77–80.
13. "Archaeological Projects," 3.
14. Interview with Gordon R. Willey; interview with Carlyle S. Smith; and Neumann, *Louisiana Archaeology*, 45.
15. "LSU–WPA;" and interview with Gordon R. Willey.

16. Willey and Sabloff, *History of American Archaeology*, 98–100, 103, 108; and WPA Monthly Reports: Louisiana, December, 1938.

17. WPA Monthly Reports: Louisiana, December, 1938; September, 1939; December, 1939; March, 1940; June, 1940; September, 1940; December, 1940; March, 1941; June, 1941.

18. The Louisiana Monthly Reports were unique to many of the WPA Reports. Extensive manuscripts accompanied these reports: Andrew Allbrecht, "Agriculture—A Survey of Ethnographic Data Pertaining to the Agriculture of the Southern Aborigines," September, 1939; James A. Ford and George I. Quimby, Jr., "Tchefuncte: A Pre-Marksville Horizon in Louisiana With an Appendix on the Skeletal Remains, by Charles E. Snow," December, 1939; Andrew Allbrecht, "Bibliography of Southeastern Ethnographic Sources," March, 1940; "Ethnographic Survey of the Aboriginal Southeast," September, 1940; Andrew C. Allbrecht "Ethno-Historical Data Pertaining to an Early Historic Indian Tribe of Louisiana—the Bayogoula," November, 1940; George I. Quimby, Jr., "European Trade Objects from Contact Period sites in the Lower Mississippi Valley," December, 1940; George I. Quimby, Jr., "The Natchezan Culture Type," March, 1941; Edwin B. Doran, "A Preliminary Survey of Archaeological Sites in Southwest Louisiana," March, 1941; Walter Beecher, "Analysis of Southwestern Louisiana Archaeological Collection," March, 1941; Carlyle S. Smith, "A Preliminary Analysis of the Pottery from the Baptiste Site (AV-25)," March, 1941; George I. Quimby, Jr., "The Tchefuncte Culture and the Subdeltas of the Mississippi River," June 1941; and George I. Quimby, "European Trade Artifacts in the Lower and Upper Mississippi Valleys," June, 1941 ; James A. Ford, "Greenhouse: A Troyville–Coles Creek Period Site in Avoyelles Parish, Louisiana," *American Museum of Natural History Anthropological Papers* 44 (1951): 32; and *American Antiquity* 1 (1935): 113; 1 (April 1936): 305; 3 (April 1938): 363; 4 (April 1939): 358; 5 (October 1939): 160: 5 (January 1940): 238. See also Albert C. Spaulding, "Fifty Years of Theory," *American Antiquity* 50 (1985): 305.

19. NA, WPA, Division of Information, Box 57, File 780-B, Louisiana.

20. James B. Griffin, *The Fort Ancient Aspect: Its Cultural And Chronological Position in Mississippi Valley Archaeology* (Ann Arbor: University of Michigan Press, 1943); "Eastern North American Archaeology: A Summary," *Science* 156 (1967): 175–91; "Culture Periods in Eastern United States Archaeology," in Griffin, *Archaeology of the Eastern United States*, 352–64; and see also James A. Ford and Gordon R. Willey, "An Interpretation of the Prehistory of the Eastern United States," *American Anthropologist* 4 (1941): 325–63.

21. Schwartz, *Conceptions*, 66–68.

22. W. S. Webb and W. D. Funkhouser, "Rock Shelters in Menifee County Kentucky," *The University of Kentucky Reports in Archaeology and Anthropology*, Vol. 3, no. 4, 1936; W.S. Webb and W.D Funkhouser, "The Ricketts Revisited, Site 3," *The University of Kentucky Reports in Archaeology and Anthropology*, Vol. 4, no. 6, 1940; W. S. Webb and W. G. Haag, "Cypress Creek Villages, Sites 11 and 12," *The University of Kentucky Reports in Archaeology and Anthropology*, Vol. 4, no. 2, 1940; W. S. Webb, "Mt. Horeb Earthworks, Site 1, and the Drake Mound, Site 11, Fayette County, Kentucky," *The University of Kentucky Reports in Archaeology and Anthropology*, Vol. 5, no. 2, 1941; W.S. Webb, "The Morgan Stone Mound Site 15, Bath County, Kentucky," *The University of Kentucky Reports in Archaeology and Anthropology*, Vol. 5, no. 3, 1941; W.S. Webb, "The C. and O. Mounds at Paintsville, Sites Jo2 and Jo9, Johnson County, Kentucky," *The University of Kentucky Reports in Archaeology and Anthropology*, Vol. 5, no. 4, 1942; W.S.

Webb, "Indian Knoll, Site Oh2, Ohio County, Kentucky," Vol. 4, no. 3, pt. 1, 1946; W.S. Webb, "The Wright Mounds, Sites 6 and 7," *The University of Kentucky Reports in Archaeology and Anthropology*, Vol. 4, no. 1, 1940; W.S. Webb, "The Riley Mound, site Be15 and the Landing Mound," *The University of Kentucky Reports in Archaeology and Anthropology*, Vol. 5, no. 3, pt. 1, 1943; W.S. Webb, "Lexman Mound, Site Be32," *The University of Kentucky Reports in Archaeology and Anthropology*, Vol. 5, no. 6, 1943; W.S. Webb,"The Carlson Annis Mound, Site 5, Butler County, Kentucky," *The University of Kentucky Reports in Archaeology and Anthropology*, Vol. 7, no. 4, 1950; W.S. Webb "The Read Shell Midden, Site 10, Butler County, Kentucky," *The University of Kentucky Reports in Archaeology and Anthropology*, Vol. 7, no. 5, 1950; W.S. Webb, The Parrish Village Site, Site 45, Hopkins County, Kentucky," *The University of Kentucky Reports in Archaeology and Anthropology*, Vol. 7, no. 6, 1951; W.S. Webb, "The Jonathan Creek Village, Site 4, Marshall County, Kentucky," *The University of Kentucky Reports in Archaeology and Anthropology*, vol. 8, no. 1, 1952; W. S. Webb and J. B. Elliott, "The Robbins Mounds, Sites Be3 and Be14, Boone County, Kentucky," *The University of Kentucky Reports in Archaeology and Anthropology*, Vol. 5, no. 5, 1942; W. S. Webb and W. G. Haag, "The Chiggerville Site, Site 1, Ohio County, Kentucky," Vol. 4, no. 1, 1939; W.S. Webb and W.G. Haag, "Archaic Sites in McLean County, Kentucky," *The University of Kentucky Reports in Archaeology and Anthropology*, Vol. 7, no. 1, 1947; W.S. Webb and W.G. Haag, "The Fisher Site, Fayette County, Kentucky," *The University of Kentucky Reports in Archaeology and Anthropology*, Vol. 7, no. 2, 1947; W. S. Webb and W. D. Funkhouser, "The Chilton Site in Henry County, Kentucky," *The University of Kentucky Reports in Archaeology and Anthropology*, Vol. 3, no. 5, 1937; and W. S. Webb and Charles E. Snow, *The Adena People, with a Chapter on Adena Pottery and a Foreword to the New Edition by James B. Griffin* (Knoxville: The University of Tennessee Press, 1974.)

23. Schwartz, *Conceptions*, 55–56.

24. Ibid., 52, 65–71.

25. "WPA Index"; WPA Monthly Reports: Kentucky, #30304, #40,311, #50,190.

26. "WPA Index"; WPA Monthly Reports: Kentucky, #30304, #40,311, #50,190.

27. Webb, "Carlson Annis Mound," 351–55; Webb, "Read Shell Midden," 355–61; Webb and Haag, "Chiggerville Site," 50–67; and Webb and Haag, "Cypress Creek Villages," 60–70. See also William A. Ritchie, "Fifty Years of Archaeology in the Northeastern United States: A Retrospect," *American Antiquity* 50 (1985): 412–20.

28. Webb, "Carlson Annis Mound," 357.

29. Webb and Haag, "Cypress Creek Villages," 66–67; and WPA Monthly Reports, #30304, August, 1938, and August, 1939.

30. Webb and Haag, "Archaic Sites in McLean County," 1–28; and WPA Monthly Reports, #30,304, April, 1938; July, 1938; December, 1938; March, 1939; June, 1939; August, 1939; December, 1939; #40,311, March, 1940; June, 1940; September, 1940; December, 1940; #50,190, June, 1941; September, 1941.

31. Schwartz, *Conceptions*, 70–74; Willey and Sabloff, *History of American Archaeology*, 130; and Haag, "Twenty Five Years," 16–19.

32. Griffin, "Adena, Foreword," v–vi; and Schwartz, *Conceptions*, 90–91.

33. Griffin, "Adena, Foreword," ix; Webb and Snow, *Adena*, 16–28, 243; and Schwartz, *Conceptions*, 95.

34. Griffin, "Adena, Foreword," x–xi; and Webb and Snow, *Adena*, 328. Another interesting interpretation about Adena culture came from Charles Snow, the physical

anthropologist who consulted with Webb. Snow was convinced that the Adena had Mesoamerican or Middle American origins. The artifacts and the pottery analyzed did not sustain that interpretation.

35. Haag, "The Pottery from the C and O Mounds at Paintsville," in Webb, "C. and O. Mounds," 341–49.

36. Webb, "Morgan Stone Mound"; Webb, "Mt. Horeb Earthworks;" and Schwartz, *Conceptions*, 95–103.

37. Schwartz, *Conceptions*, 66–74.

38. Ibid., 106–109; and James B. Griffin, "The Chronological Position and Ethnological Relationships of the Fort Ancient Aspect," *American Antiquity*, 2 (1937): 273–76.

39. Schwartz, *Conceptions*, 109–15; and Webb, "Jonathan Creek Village"; and *American Antiquity* 1 (1935): 113; 1 (1936): 329; 3 (1937): 192; 5 (1939): 160; 5 (1940): 237; 6 (1941): 351.

40. Schwartz, *Conceptions*, 116–18; interview with James B. Griffin; Haag, "Twenty Five Years," 16–19; Stoltman, "Southeastern United States," 136–42; and Jesse D. Jennings, "Prehistory of the Lower Mississippi Valley," in Griffin, *Archaeology of the Eastern United States*, 256–71.

41. Schwartz, *Conceptions*, 61, 71, 113.

42. Ibid.; and T. M. N. Lewis and Madeline Kneberg, "The Prehistory of the Chickamauga Basin in Tennessee," *Tennessee Anthropology Papers* (1941). See also University of Memphis, Oral History Research Office, "Oral History of the Tennessee Valley Authority: Interviews with Thomas M. N. Lewis and Mrs. Madeline Kneberg Lewis, December 19, 1972," by Charles W. Crawford.

43. WPA, Monthly Reports: Tennessee, #20050; WPA, Division of Women and Professional, Box 57, file 780-B, reprint, *Journal of the Tennessee Academy of Sciences* 10 (1935): 153–59.

44. NAA, Matthew W. Stirling Papers, National Resources Committee File, Annual Reports of the Committee on Archaeology of the Tennessee Valley, 1935, 1936, 1937, 1938, 1939, 1940, 1941.

45. Webb Papers, Boxes 2 and 3, letter Lewis to Webb (6 February 1936), letter Webb to Lewis (19 November 1936), letter Lewis to Webb (22 November 1936), letter Webb to Lewis (10 April 1937).

46. Webb Papers, Box 5, letter Webb to Guthe (3 October 1939). Edwin A. Lyon spends some time developing the exact chronology of the feud. See Lyon, "New Deal Archaeology in the Southeast," 136–56.

47. NA, Records of the National Park Service, Record Group 79, Director's Office Files, Cammerer, 1923–40, Personal and Confidential Letters and Miscellaneous, letter Lehman to Draper (31 May 1940); and Webb Papers, Box 3, letter Kelly to Webb (22 March 1939), letter Setzler to Webb (17 March 1939), letter Webb to Fechner (30 March 1939), letter Fechner to Webb (3 April 1939), letter Webb to Wirth (5 April 1939), letter Johnston to Webb (13 April 1939). Even the Tennessee Valley Authority, in an economy move, pressured Lewis to draw his labor supply from the Civilian Conservation Corps rather than from the Works Progress Administration.

48. "WPA Index"; WPA Monthly Reports: Tennessee, #20,050, #30,295, #40,225, #15,045, September, 1938; December, 1938; March, 1939; June, 1939; March, 1940; September, 1940, December, 1940; March, 1941; June, 1941; September, 1941.

49. WPA Monthly Reports: Tennessee, #40,225, June, 1941; and manuscript, T. M.

N. Lewis and Madeline D. Kneberg, "Prehistory of the Chickamauga Basin in Tennessee," (June 1941); and *American Antiquity* 1 (1935): 131; 1 (1936): 304, 328; 2 (1937): 314–15; 3 (1937): 193; 2 (1938): 362; 6 (1941): 351–54.

50. Webb Papers, Box 5, letter Webb to Guthe (3 July 1936); Box 4, letter Webb to Deignan (13 October 1941), letter Deignan to Webb (30 October 1941), letter Webb to Deignan (1 November 1941).

51. Webb Papers, Box 5, letter Webb to Guthe (3 July 1936); Box 4, letter Webb to Deignan (13 October 1941), letter Deignan to Webb (30 October 1941), letter Webb to Deignan (1 November 1941).

52. Webb Papers, Box 3, letter Lewis to Webb (22 November 1936), and letter Webb to Lewis (10 April 1937). See also letter Webb to Deignan (13 October 1941), letter Deignan to Webb (30 October 1941), and letter Webb to Deignan (1 November 1941) for additional perspective.

53. Webb Papers, Box 5, letter Setzler to Webb (8 October 1941), letter Webb to Setzler (13 October 1941), letter Webb to Setzler (15 April 1942), letter Setzler to Webb (18 April 1942), letter Webb to Setzler (20 April 1942), letter Strong to Webb (31 January 1942), letter Deignan to Strong (28 January 1942).

54. Webb Papers, Box 5, letter Webb to Judd (2 November 1938), letter Webb to Setzler (4 November 1938), letter Setzler to Webb (12 November 1938), letter Webb to Setzler (16 November 1938).

55. Webb Papers, Box 5, letter Jennings to Webb (6 December 1936), letter Lewis to Webb (3 September 1936), letter Webb to Jennings (10 December 1936). See also Jennings, *Accidental Archaeologist,* 84–89. Jennings relates that Lewis was not an easy person with whom to work.

56. Webb Papers, Box 5, letter Lewis to Webb (11 January 1936), letter Webb to Lewis (7 February 1936), letter Lewis to Webb (22 November 1936), letter Lewis to Webb (8 January 1937), letter Lewis to Barnes (2 February 1937).

57. Lyon, "New Deal Archaeology in the Southeast," 134–68, 196–218. The chronicle of the events follows closely but the accompanying analysis is substantively weak.

58. Strong, "Recommendations of the Committee."

59. John Dean Minton, *The New Deal in Tennessee: 1932–1938* (New York: Garland Publishing, Inc., 1979), 183–210; Madeline Kneberg, "The Tennessee Area," in Griffin, *Archaeology of the Eastern United States,* 190–98; Chandler W. Rowe, "Woodland Cultures of Eastern Tennessee," in Griffin, *Archaeology of the Eastern United States,* 199–206; and Andrew H. Whiteford, "A Frame of Reference for the Archaeology of Eastern Tennessee," in Griffin, *Archaeology of the Eastern United States,* 207–25.

60. Interview with James B. Griffin.

61. *American Antiquity* 1 (1935): 47–49.

62. Ibid.

63. William S. Webb and David L. DeJarnette, "An Archaeological Survey of Pickwick Basin in the Adjacent Portions of the States of Alabama, Mississippi and Tennessee," *Bureau of American Ethnology Bulletin* 129 (1942).

64. "WPA Index"; WPA Monthly Reports: Alabama, #20,314, #30,295, #61,002.

65. David L. DeJarnette, "Alabama Archaeology: A Summary," in Griffin, *Archaeology of the Eastern United States*; Haag, "Twenty Five years"; WPA, Monthly Reports: Tennessee, November, 1936; April, 1938; June, 1938; September, 1938; December, 1938; March, 1939; June, 1939; September, 1939; December, 1939; March, 1940;

June, 1940; September, 1940; December, 1940; March, 1941, June, 1941; September, 1941; Quimby, "Brief History," 115; NA, WPA, Information Division, Primary File: 780-B Archaeological Excavations, "Archaeological Projects Seek Light on Early American," 1-3; and NA, WPA, Information Division, Primary File: 780-B Archaeological Excavations, "Turning the Lost Pages of Prehistory," 1-3.

66. WPA Monthly Reports: Alabama, #20,314, #30,295, #61,002; and *The Geological Survey of Alabama* published six articles between 1941 and 1948.

67. FMS, Series 1, Subseries 1, Box 3, Files: 1936-38; Webb Papers, Box 4, Files 1939-40; and Webb Papers, Subseries 2, Box 10, file: Alabama.

68. FMS, Series 1, Subseries 1, Box 3, Files: 1936-38; Webb Papers, Box 4, Files 1939-40; and Webb Papers, Subseries 2, Box 10, file: Alabama.

69. Webb Papers, Box 10, letter DeJarnette to Webb (7 May 1938).

70. Ibid. James B. Griffin, "Formation of the Society," 267. Susan Ware deals with New Deal women and their active network in her excellent work, *Beyond Suffrage: Women in the New Deal* (Cambridge: Harvard University Press, 1981). Beginning with Eleanor Roosevelt and filtering down through the various administrators, e.g., Frances Perkins, Ellen Woodward, and Florence Kerr, Dr. Ware carefully documents their sensitivities toward relief, human needs, equal rights, and academic interests. See also William Henry Chafe, *The American Women: Her Changing Social, Economic, and Political Roles, 1920-1970* (New York: Oxford University Press, 1972); McJimsey, *Harry Hopkins*, 44-61; and Thomas A. Krueger and William Glidden, "The New Deal Intellectual Elite: A Collective Portrait," in *The Rich, the Well-born, and the Powerful: Elites and Upper Classes in History*, ed. Frederic Cole Jaher (Urbana: University of Illinois Press, 1973), 338-74.

71. FMS, Series 1, Subseries 1, Box 3, Files: 1936-38; Webb Papers, Box 4, Files 1939-40; and Webb Papers, Subseries 2, Box 10, file: Alabama.

72. Webb Papers, Box 3, letter DeJarnette to Webb (24 February 1936) with addendum, letter DeJarnette to Webb (26 March 1936), letter Webb to DeJarnette (30 march 1936), letter DeJarnette to Webb (1 April 1936), letter Webb to DeJarnette (3 April 1936), letter DeJarnette to Webb (7 April 1936), telegram DeJarnette to Webb (1 May 1936), letter Webb to DeJarnette (2 May 1936).

73. Webb Papers, Box 3, letter Webb to DeJarnette (2 May 1936), letter Jones to Levron (23 January 1937), letter Jones to Webb (29 January 1937), letter Webb to Jones (29 January 1937), letter Webb to Neely (23 March 1937).

74. WPA, Division of Women and Professional, Box 57, Monthly Reports: Alabama, September, 1938.

75. Webb Papers, Box 3, letter Russell to Lee (23 March 1938); and FMS, Series 1, Subseries 2, Box 10, letter Kelly to Setzler (7 October 1938).

76. Webb Papers, Box 3, letter Jones to Petrullo (29 July 1938), letter Webb to Haygood (26 March 1938), letter Webb to Petrullo (5 August 1938), telegram Petrullo to Webb (15 October 1938).

77. WPA, Division of Information, box 57, File 780-B.

78. WPA Monthly Reports: Alabama; and interview with James B. Griffin.

79. Webb Papers, Box 3, letter Webb to Deignan (13 October 1941), letter Deignan to Webb (30 October 1941), letter Webb to Deignan (1 November 1941), letter Deignan to Webb (n.d.), letter Setzler to Webb (8 October 1941), letter Webb to Setzler (15 April 1942), letter Setzler to Webb (18 April 1942), letter Webb to Setzler (20 April

1942), letter Strong to Webb (31 January 1942), letter Deignan to Strong (28 January 1942), letter Webb to Storey [sic] (2 February 1942).

80. Webb Papers, Box 3, letter Webb to Deignan (13 October 1941), letter Deignan to Webb (30 October 1941), letter Webb to Deignan (1 November 1941), letter Deignan to Webb (n.d.), letter Setzler to Webb (8 October 1941), letter Webb to Setzler (15 April 1942), letter Setzler to Webb (18 April 1942), letter Webb to Setzler (20 April 1942), letter Strong to Webb (31 January 1942), letter Deignan to Strong (28 January 1942), letter Webb to Storey [sic] (2 February 1942); and FMS, Series 1, Subseries 2, Box 5, 1941, File: J. D. Jennings—correspondence on missing Peachtree documentary material and evaluation of WPA personnel in Alabama.

81. Webb and DeJarnette, "Survey of Pickwick Basin," 2–7.

82. Ibid., 301–21.

83. Ibid., 159–78, 92–93.

84. Ibid., 509–26.

85. Ibid.

86. WPA Monthly Reports: Alabama, June, 1938; September, 1938; December, 1938; March, 1939; June, 1939; September, 1939; December, 1939; March, 1940; and W. S. Webb and Charles G. Wilder, *An Archaeological Survey of Guntersville Basin on the Tennessee River in Northern Alabama* (Lexington: University of Kentucky Press, 1951), 1–8.

87. Webb and Wilder, *Guntersville Basin*, 266–72.

88. Webb Papers, Box 3, letter DeJarnette to Webb (4 May 1938). The efforts of black workers on digs are explored in further detail in the section on Georgia, where the phenomenon was more extensive.

89. WPA Monthly Reports: Alabama, March, 1941; June, 1941; September, 1941; December, 1941; March, 1942; April, 1942; June, 1942.

90. Ibid.; Steve B. Wimberly and Harry A. Tourtelot, "The McQuorquodale Mound: A Manifestation of the Hopewellian Phase in South Alabama," *Geological Survey of Alabama*, no. 19, 1941; David L. DeJarnette and Steve B. Wimberly, "The Bessemer Site: Excavation of Three Mounds and Surrounding Village Areas near Bessemer, Alabama," *Geological Survey of Alabama*, no. 17, 1941; Charles E. Snow, "Condylo-Diaphysial Angles of Indian Humeri from North Alabama," *Geological Survey of Alabama*, no. 16, 1940; "Two prehistoric Indian Dwarf Skeletons from Moundville," *Geological Survey of Alabama*, no. 21, 1943; W. S. Webb and D. L. DeJarnette, "Little Bear Creek Site Ct°8, Colbert County, Alabama," *Geological Survey of Alabama*, no. 26, 1948; "The Perry Site Lu°25, Units 3 and 4, Lauderdale County, Alabama," *Geological Survey of Alabama*, no. 25, 1948; and *American Antiquity* 1 (July 1935): 47; 6 (1941): 3; 6 (1941): 352.

91. WPA, Division of Information, Box 56, 780-A, Box 2, 780-B; and Division of Women and Professional, Box 57, 780-B.

92. Michael S. Holmes, *New Deal in Georgia: an Administrative History* (Westport, CT: Greenwood Press, 1975), 61–95; McJimsey, *Harry Hopkins*, 61–63. McJimsey also details a similar situation in Louisiana, where the WPA had to work against the power of Huey Long. The net result was the same: the creation of a parallel relief organization (p. 92).

93. FMS, Series 1, Subseries 2, Boxes 10–12, Georgia.

94. FMS, Series 1, Subseries 2, Box 11, letter Kelly to Fewkes (4 January 1937).

95. FMS, Series 1, Subseries 2, Box 11, letter Kelly to Setzler (9 December 1937).

96. FMS, Series 1, Subseries 2, Box 10, letter Kelly to Setzler (2 June 1937), letter Kelly to Setzler (10 June 1937).

97. Gordon R. Willey, "The Social Uses of Archaeology," *The Kenneth B. Murdock Lecture*, Leverett House, Harvard University, March, 1980; interview with Gordon R. Willey; Webb Papers, Box 3, letter Kelly to DeJarnette (14 February 1938), letter Ford to DeJarnette (14 February 1938), letter Ford to DeJarnette (17 February 1938), letter Webb to DeJarnette (28 February 1938); and Jennings, *Accidental Archaeologist*, 103–13.

98. FMS, Series 1, Subseries 2, Box 10, letter Setzler to Kelly (21 December 1937), letter Kelly to Setzler (8 January 1938).

99. WPA, Division of Information, box 57, File 780-B.

100. "WPA Index"; WPA Monthly Reports: Georgia, #30,029, September, 1938; December, 1939; March, 1940; June, 1940; September, 1940; December, 1940.

101. Arthur R. Kelly, "A Preliminary Report on Archaeological Explorations at Macon, Georgia," *Bureau of American Ethnology Bulletin* 119 (1938).

102. Ibid., 2.

103. WPA Monthly Reports: Georgia, #20,347, June, 1936; August, 1936; October, 1936; December, 1936; March, 1937; July, 1937; and FMS, Series 1, Subseries 2, box 11, letter Kelly to Setzler (9 December 1937).

104. WPA Monthly Reports: Georgia, #20,420, January, 1938.

105. WPA Monthly Reports: Georgia, #20,429, January, 1938, February, 1938; June, 1938; July, 1939; August, 1938; January, 1939; April, 1939; October, 1939; January, 1940; August, 1940; December, 1940; "American Prehistory," 56–57; "Turning Lost Pages of History," 1–4; "Archaeological Projects Seek Light," 5–7; and Quimby, "Brief History," 113–14.

106. Interview with Gordon R. Willey; and WPA, Division of Women and Professional, Box 57, 780-B, #13960. For a further discussion of blacks in the New Deal, see "Part Nine: The Negro," in Howard Zinn, ed., *New Deal Thought* (Indianapolis: The Bobbs-Merril Co., Inc., 1966); and Robert C. Weaver, "The Black Cabinet," in *The Making of the New Deal: The Insiders Speak*, ed.Katie Louchheim (Cambridge: Harvard University Press, 1983), 261–64.

107. WPA, Division of Women and Professional, Box 57, 780-B, #13859, Vladimir J. Fewkes, "W.P.A. Excavations at Irene Mound," 8.

108. WPA, Division of Information, box 57, File 780-B, text of speech by Lucy McIntyre before Society for Georgia Archaeology, Athens, GA, 14 October 1938, 1–4.

109. "WPA Index"; WPA Monthly Reports: Georgia, #75,006, December, 1938; March, 1939; June, 1939; September, 1939, December, 1939; March, 1940, June, 1940; September, 1940.

110. "WPA Index"; WPA Monthly Reports: Georgia, #75,006, December, 1938; March, 1939; June, 1939; September, 1939, December, 1939; March, 1940, June, 1940; September, 1940; Robert Wauchope, "Archaeological Survey of Northern Georgia," *Memoirs of the Society for American Archaeology* 21 (1966); and *American Antiquity* 2 (1936): 72; 3 (1938): 362; 6 (1940): 352.

CONCLUSION

1. For an extended discussion of the political activities of archaeologists in this formative period, see Gregory Jon Marlowe, "Preservation and Profession in American Ar-

chaeology: From 'Basic Needs' to the Committee for the Recovery of Archaeological Remains, 1939–1945," paper presented at the Pacific Coast Branch, American Historical Association, Annual Meeting, San Diego, 11 August 1983. Additional narratives of the political background and archaeological concerns are included in Carl Guthe, "The Basic Needs of American Archaeology," *Science* 90 (1939): 528–30; Frederick Johnson, Emil Haury, and James B. Griffin, "The Planning Committee of the Society for American Archaeology: A Preliminary Statement," *American Antiquity* 10 (1945): 220; James B. Griffin, "Report of the Planning Committee," *American Antiquity* 12 (1945): 142–44; Frederick Johnson, "The Work of the Committee for the Recovery of Archaeological Remains: Aims, History, and Activities to Date," *American Antiquity* 13 (1947): 212–15; Frederick Johnson, "Archeology in an Emergency: The federal government's Inter-Agency Archaeological Salvage Program is 20 years old," *Science* 152 (1966): 1592–97; and Frederick Johnson, "The Inter-Agency Archaeological Salvage Program in the United States," *Archaeology* 4 (1951): 25–40; see also The Committee For The Recovery Of Archaeological Remains, *The Inter-Agency Archaeological Salvage Program: After Twelve Years* (Columbia: University of Missouri Press, 1958), 1–3.

2. Guthe, "The Basic Needs of American Archaeology," 528–30.

3. Ibid.; and Johnson, "Inter-Agency," 25–27.

4. Guthe, "The Basic Needs of American Archaeology," 528–30; Johnson, "Inter-Agency," 25–27; and Johnson, "Archeology in an Emergency," 1594–95.

5. Guthe, "The Basic Needs of American Archaeology," 528–30; Johnson, "Inter-Agency," 25–27.

6. Guthe, "The Basic Needs of American Archaeology," 528–30; Johnson, "Inter-Agency," 25–27.

7. Guthe, "The Basic Needs of American Archaeology," 528–530; Johnson, "Inter-Agency," 25–27.

8. Interview with Frederick Johnson, Boston, MA, August, 1983.

9. Johnson, "Archeology in an Emergency," 1594.

10. Frederick Johnson, et al., "The Planning Committee of the Society for American Archaeology," 220.

11. Frederick Johnson, et al., "Report of the Planning Committee," 142.

12. Ibid.; and Johnson, "The Work of the Committee," 212–13.

13. Johnson, et al., "Report of the Planning Committee," 142; Johnson, "Work of the Committee," 212–13; and Herbert E. Kahler, "The Role of the National Park Service in River Basin Archaeology with Particular Reference to Inter-Bureau Agreements," *American Antiquity* 13 (1947): 215–16.

14. Johnson, "The Work of the Committee," 213.

15. Interview with James B. Griffin. It must be understood, Griffin emphasized that the plan to survey the Central Mississippi Valley was the seed idea of three men. Consequently, this was *not* a disciplinewide notion.

16. NA, Record Group 70, Records of the National Park Service, Central Classified File, 1933–49, General Files: 740–02, Archaeology, "Plan For An Archaeological Survey Of The Central Mississippi Valley, Submitted to the U.S. National Park Service by James B. Griffin, Phillip Phillips and James A. Ford," 1–11.

17. Ibid.

18. Ibid.; "Proposed Budget For A Survey Of The Archaeology Of The Mississippi Valley, 2nd Proposal," 1–2.

19. Interview with James B. Griffin.

Epilogue

1. Taylor, *A Study of Archaeology*.
2. Ibid., 24–39.
3. Ibid., 43–46.
4. Ibid., 46–66.
5. Ibid., 66–87; and Willey and Sabloff, *American Archaeology*, 160–64.
6. Taylor, *A Study of Archaeology*, 150–200.
7. Haury, "Reflections," 388. Dr. Haury further prefaced these remarks by stating: "In the later years, I remember hearing criticisms of the archaeologists of that time for being too entranced with the data and the collecting of it, but not doing enough about extracting the meaning. In their defense, I must say that the gathering of information is where the study begins and forever remains the touchstone to good archaeology. I further hold that a kind of critical mass of data must be reached before successful efforts can be made to extract the essence" (p. 387). In short, each generation builds on the work of those that precede it.

Perhaps the controversy over "New Archaeology," viewed in the larger scenario of historical professional development, is neither negative nor positive; rather, it is an indication of the further changes occurring in the discipline. Debate and controversy stand as a healthy sign of progress. New perspectives seek their own identity, much as the older profession did.

8. Guthe, "Reflections On The Founding," 439–40.
9. Haag, "Twenty-Five Years of Eastern Archaeology, 19.
10. Ibid., 21–22.

Appendix 1

1. American Antiquity 1 (1935): 113; and "American Prehistory," 23–24.
2. *American Antiquity* 1 (1935): 113; and James A. Ford and George I. Quimby, Jr., "The Tchefuncte Culture: An Early Occupation of the Lower Mississippi Valley," *Memoirs of the Society for American Archaeology* 2 (1945): 24.
3. NAA, Ford Papers, Box: Louisiana.
4. Smith, "History of the Society," 15–16.
5. *American Antiquity* 1 (1935): 115–16.
6. Ibid., 116–17; and "American Prehistory," 25.
7. "American Prehistory," 24–25.
8. Ibid.; and *American Antiquity* 1 (1935): 120–23.
9. *American Antiquity* 1 (1935): 122.
10. Ibid., 119–20.
11. Ibid., 128.
12. Ibid., 135–37.
13. Ibid., 133–34.
14. Ibid., 124–25. See also. Stocking, "Ideas and Institutions," 12. Philanthropic support was especially critical in the early Depression.

15. *American Antiquity*, 1 (1935): 117–18.
16. *American Antiquity*, 1 (1935): 121.
17. Interview with Emil W. Haury, University of Tucson, Tucson, Arizona, October 1982.
18. Harold S. Gladwin and Winifred Gladwin, *A Method for the Designation of Cultures and Their Variations* (Pasadena: Medallion Papers, 1928); *American Antiquity* 1 (1935): 124; and Willey and Sabloff, *American Archaeology*, 106–107, 122–23.
19. *American Antiquity*, 1 (1935): 118–19.
20. Interview with Emil W. Haury.

APPENDIX 2

1. Interview with Emil W. Haury; "WPA Index," Arizona; WPA Monthly Reports: Arizona, #30,002, December, 1938; March, 1939; June, 1939; September, 1939; December, 1939; March, 1940; June, 1940. See also Jimmy H. Miller, *The Life of Harold Sellers Colton: A Philadelphia Brahmin in Flagstaff* (Tasaile, AZ: Navajo Community College Press, 1991).
2. Interview with Emil W. Haury; WPA, Division of Women and Professional, 780-B, "Digest of WPA Archaeological Projects," File #13699; "American Prehistory," 32–33; Robert H. Lister, "Twenty Five Years of Archaeology in the Greater Southwest," *American Antiquity* 27 (1961): 39–45; 1 (1935): 98–108.
3. *American Antiquity* 2 (1937): 217.
4. *American Antiquity* 2 (1936): 50–52; 2 (1936): 146; 2 (1937): 264–66; 2 (1937): 282–88; 3 (1937): 3–33; 3 (1937): 353; 5 (1938): 68–69, 71; 4 (1939): 353; 5 (1939): 69–70; 5 (1939): 161–62; 5 (1940): 249; 5 (1940): 343; 6 (1940): 86–87; 6 (1940): 176–77; 6 (1941): 282–83; 7 (1941): 79; 7 (1941): 183.
5. WPA, Division of Information, Box 56, File 780-A, press releases July 1935 and 4 April 1939.
6. *American Antiquity* 1 (1935): 120–21; 2 (1936): 48–49; 3 (1938): 274; 4 (1938): 70; and Carol A. Gifford and Elizabeth A. Morris, "Digging For Credit: Early Archaeological Field Schools In The American Southwest," *American Antiquity* 50 (1985): 395–411. See also Beatrice Chauvenet, *Hewett and Friends: A Biography of Santa Fe's Vibrant Era* (Santa Fe: Museum of New Mexico Press, 1983).
7. *American Antiquity* 5 (1940): 247–51; 6 (1940): 84–85; 6 (1941): 281; 6 (1941): 356; 7 (1941): 78; 7 (1941): 182; WPA, Division of Women and Professional, File 780-B, #13612, #13699; SIA, Secretary, letter Woodward to Stirling (20 October 1937); letter Stewart Peckham to author (5 December 1985) and accompanying memorandum, "Anthropology Faculty, Visiting Instructors, Graduate Fellows and Students Assistants at Univ. of New Mexico from 1928 through 1948," prepared for Albuquerque Branch of Cultural Research, National Park Service, Department of Interior; and letter Dr. Frances Joan Mathien to author (23 December 1985) and accompanying memorandum, "Participation of the Museum of New Mexico and affiliated institutions in Federal Works Programs—1935-1943," Ewing to Peckham (6 January 1975).
8. NAA, Records of the BAE, Additional Admin. Files, 1926–46, letter Petrullo to Stirling (3 May 1938).

9. Ibid.

10. Ibid.; letter Mekeel to Stirling (2 May 1938); and see Warren O. Hagstrom, "Competition in Science," *American Sociological Review* 39 (1974): 1–18.

11. NAA, Records of the BAE, Additional Admin. Files, 1926–46, memorandum Stirling to Abbot (4 May 1938); letter Hill to Stirling (8 April 1938). George W. Stocking, Jr., precisely addresses the regional orientation of Hewett in his article, "The Santa Fe Style in American Anthropology: Regional Interest, Academic Initiative, and Philanthropic in the First Two Decades of the Laboratory of Anthropology, Inc.," *Journal of the History of the Behavioral Sciences* 18 (1982): 3–19. He begins by generalizing: "The conflict between scientific universalism and localized particularistic concerns is a theme recurring at many moments and different levels in the history of anthropology," both in contests of research orientations (e.g., progressive evolutionism) and between institutional entities. The specific issue here is the matter of an alternative research entity to circumvent the tenacity of one dominating individual: Hewett. The focus of this study is how Hewett's antagonism turned toward the larger sphere the federal government. Stocking also makes the point that many of the new breed of anthropologists and archaeologists were trained by Franz Boas or his students. Hewett's training dates from a time and group adamantly opposed to Boas.

12. James Denney, "History Observes a Centennial," in *Magazine of the Midlands*, Sunday Edition, *The Omaha World-Herald*, 24 September 1978, 19–22.

13. *American Antiquity* 1 (1935): 117–18.

14. Interview with Carlyle S. Smith; and interview with Marvin Kivett.

15. Interview with Carlyle W. Smith; and interview with Marvin Kivett.

16. "Archaeological Projects Seek Light on Early American," 1–4. The files of the Information Division contained hundreds of short articles from local newspapers. Most were only two- or three-inch columns, while a few others were front page size. The point made here is that the WPA effectively and consistently prompted media coverage.

17. Denney, "Centennial": W. D. Strong, et al., "Asa T. Hill," *Nebraska History Magazine* 34 (1953): 67–90; interview with Marvin Kivett; and interview with Carlyle S. Smith.

18. Interview with Carlyle S. Smith.

19. Ibid.

20. Denney, "Centennial"; and interview with Marvin Kivett.

21. "WPA Index"; WPA Monthly Reports: Nebraska, #20,337, #50,005, #30,310, December, 1938; September, 1939; December, 1939; December, 1940; March, 1941; September, 1941; and *American Antiquity* 2 (1937): 231.

22. "WPA Index"; WPA Monthly Reports: Nebraska, #20,321, #30,310, #50,005, September, 1938; December, 1938; September, 1939; December, 1939; March, 1941; September, 1941; "American Prehistory," 61–62; George C. Frison, "The Plains," in Fitting, *Development*, 160–62; "Archaeological Projects Seek Light" 3; *American Antiquity* 2 (1936): 52; 2 (1936): 146; 2 (1937): 309–11; 3 (1938): 275–76; 4 (1939): 282; 5 (1939): 68; 5 (1940): 243–44; Nebraska State Historical Society, A. T. Hill Collection, Box 6, File 98, "Brief," prepared by A. E. Sheldon; A. T. Hill and Waldo R. Wedel, "Excavations at the Leary Indian Village and Burial Site, Richardson County, Nebraska, *Nebraska History Magazine* 17 (1936): 3–6; A. T. Hill and Paul Cooper, "The Schrader Site," *Nebraska History Magazine* 17 (1936): 223–25; A.T. Hill and Paul Cooper,"The Champe Site," *Nebraska History Magazine* 18 (1937): 253–54; A.T. Hill and Paul Cooper,

"The Archaeological Campaign of 1937 by the Nebraska State Historical Society," *Nebraska History Magazine* 18 (1937): 243–47; Paul Cooper, "Report of Explorations," *Nebraska History Magazine* 20 (1939): 95–99; A. E. Sheldon, "Foreword," and "Introduction," *Nebraska History Magazine* 21 (1940): 143–45, 147–52; A.E. Sheldon, "Introduction," *Nebraska History Magazine* 22 (1941): 159–61.

23. "WPA Index"; WPA Monthly Reports: Nebraska, #20,338, #30,243, #50,005, September, 1938; December, 1938; March, 1941; September, 1941; Carlyle S. Smith Personal Papers, "The Objectives and Methods of Project 5517," Summer, 1939; interview with Carlyle Smith; interview with Marvin Kivett; NA, WPA Files, Division of Information, Box 56, File 780-B, "American Prehistory," 61; *American Antiquity* 2 (1937): 222–23; 3 (1937): 191; 3 (1937): 276–78; 4 (1939): 355–57; 5 (1940): 242–44.

24. *American Antiquity* 2 (1937): 221; and A. T. Hill Collection, Box 1, File 13, memorandum from John L. Champe.

25. *American Antiquity* 3 (1937): 275–76.

26. Nebraska State Historical Society, John L. Champe Collection, Box 1; and interview with Marvin Kivett.

27. Earl F. Bell, *Chapters in Nebraska Archaeology* (Omaha: University of Nebraska Press, 1936); and Waldo R. Wedel, "An Introduction to Pawnee Archaeology," *Bureau of American Ethnology Bulletin* 112 (1936).

28. "WPA Index"; WPA Monthly Reports: Pennsylvania, #20,169, #30,917, #40,592, #23,676, #19,421, #18,369, #23,174, #40,250, #22,831; October, 1938; December, 1938; March, 1939; June, 1939; September, 1939; March, 1940; June, 1940; September, 1940.

29. "WPA Index"; WPA Monthly Reports: Pennsylvania, #20,169, #30,917, #40,592, #23,676, #19,421, #18,369, #23,174, #40,250, #22,831; October, 1938; December, 1938; March, 1939; June, 1939; September, 1939; March, 1940; June, 1940; September, 1940.

30. "WPA Index"; WPA Monthly Reports: Pennsylvania, #20,169, #30,917, #40,592, #23,676, #19,421, #18,369, #23,174, #40,250, #22,831; October, 1938; December, 1938; March, 1939; June, 1939; September, 1939; March, 1940; June, 1940; September, 1940; and NA, WPA Files, Division of Information, Box 56, File 780-B, "American Prehistory," 57–60.

31. WPA Monthly Reports: Pennsylvania, #32,337, #41,319, #75,155; March, 1939; June, 1939; September, 1939; December, 1939; January, 1940; March, 1940; June, 1940; WPA, Division of Women and Professional, Box 57, File 1938, 17; WPA, Information Files, Box 57, File 780-B, 3; SIA, Secretary, letter Woodward to Stirling (20 October 1937); *American Antiquity* 1 (1935): 129–30; 2 (1936): 55; 2 (1937): 313–14; 3 (1937): 86–87; 5 (1939): 59; and Edgar E. Augustine, "Recent Discoveries in Somerset County," *Pennsylvania Archaeologist* 8 (1938): 6–12.

32. *American Antiquity* 3 (1937): 85; 3 (1938): 279; 4 (1939): 357–58; 5 (1939): 157; 5 (1940): 205, 234; 5 (1940): 340.

33. *American Antiquity* 3 (1937): 85; 3 (1938): 279; 4 (1939): 357–58; 5 (1939): 157; 5 (1940): 205, 234; 5 (1940): 340.

34. Interview with Marvin Kivett; and interview with Carlyle S. Smith.

35. "WPA Index"; WPA Monthly Reports: Wisconsin, #30,185, February, 1938; and "Archaeological Projects Seek Light," 4.

36. *American Antiquity* 1 (1935): 138.

37. *American Antiquity*, 2 (1936): 53; 2 (1936): 147–48.
38. *American Antiquity*, 1 (1935): 64–65, 138; 2 (1936): 53; 2 (1936): 147–48.
39. "WPA Index"; WPA Monthly Reports: Indiana, #30,392, May, 1940; September, 1940; March, 1941; September, 1941; *American Antiquity* 5 (1939): 166–67; 5 (1940): 240; 6 (1940): 83; 6 (1940): 176; 7 (1941): 181; James H. Kellar, "Glenn A. Black, 1900–1964," *American Antiquity* 31 (1966): 402–405; "Glenn A. Black," *Indiana Magazine of History* 63 (1967): 49–52; and letter Kellar to author (26 November 1985). Dr. Kellar graciously supplied the information on Professor Black.
40. *American Antiquity* 1 (1935): 63–64; 3 (1937): 192; 5 (1939): 165–66; 5 (1940): 239–40.
41. *American Antiquity* 5 (1939): 165–66; 5 (1940): 239–40.
42. *American Antiquity* 3 (1937): 192.
43. *American Antiquity* 2 (1937): 224–25; 5 (1939): 165–66; 5 (1940): 239–40; 6 (1940): 82; 6 (1940): 277, 279–80.
44. *American Antiquity* 1 (1935): 63–64; 5 (1939): 165–66; 5 (1940): 239–40; 6 (1940): 82; 6 (1941): 279–80; 7 (1941): 76–77; 7 (1941): 81; and WPA, Division of Information, Box 56, File 780-A, press release 23 January 1941.
45. "WPA Index"; WPA Monthly Reports: Montana, #10513, September, 1937; September, 1938; December, 1938; #40,117, December, 1940; March, 1941; June, 1941; September, 1941; October, 1941; NA, WPA, Information Files, "WPA Archaeological Projects," Box 57, File 780-B, 10 Nov 1938, 2; "Petrullo Interview"; NA, WPA, Monthly Report Montana, September, 1938; and WPA, Division of Women and Professional, Box 57, File 780-B, #13612.
46. *American Antiquity* 1 (1935): 117; 2 (1936): 59; 3 (1938): 190, 278; "WPA Index"; and WPA, Division of Information, Box 56, File 780-A, press release 24 August 1941.
47. "WPA Index," Wyoming; WPA Monthly Reports: Wyoming, #30,131, December, 1938; March, 1939; September, 1939; December, 1939; March, 1940; June, 1940; September, 1940; December, 1940; March, 1941; April, 1941; "Petrullo Interview"; and "American Prehistory," 63.
48. WPA Index, WPA Monthly Reports, Wyoming, Final Report, #30,131, March, 1941; and *American Antiquity* 2 (1936): 60.
49. "Petrullo Interview," 63.
50. NAA, "WPA Index," California; and *American Antiquity* 1 (1935): 52; 1 (1935): 119; 1 (1936): 295–300; 2 (1936): 47, 59; 2 (1936): 108–16; 2 (1937): 69, 137–41; 4 (1939): 280, 287; 5 (1939): 68–69; 5 (1940): 252–53; 5 (1940): 346; 6 (1940): 175; 6 (1941): 214–16, 284–85; 7 (1941): 80; 7 (1941): 98–122, 123–33, 134–46; 7 (1942): 184–85.
51. "WPA Index"; WPA Monthly Reports: California, #85,076 and #30,397, July, 1937; October, 1938; December, 1938; April, 1939; January, 1940; "American Prehistory," 53; and FMS, WPA Correspondence, Box 10, California, letter Wetmore to Setzler (7 December 1938).
52. "American Prehistory," 56.
53. Roderick Sprague, "The Pacific Northwest," in Fitting, *Development*, 258–61; Clement W. Meighan, "The Growth of Archaeology in the West Coast and the Great Basin, 1935–60," *American Antiquity* 27 (1961): 33–34; and *American Antiquity* 1 (1935): 129; 2 (1936): 47; 3 (1937): 53, 67, 82; 3 (1938): 272–73; 4 (1939): 280–81; 5 (1940): 346–47; 6 (1940): 19–28; 6 (1941): 339–42; 7 (1941): 147–55; 7 (1942): 414.

54. SIA, Secretary, letter Woodward to Stirling (20 October 1937); Sprague, "Northwest," 251–85; *American Antiquity* 5 (1940): 251; 6 (1940): 140; 6 (1941): 359; and "Archaeological Projects Seek Light."

55. *American Antiquity* 1 (1935): 55; 1 (1935): 155; 2 (1936): 57; 2 (1936): 217, 220; 2 (1937): 316; 3 (1937): 190; 3 (1938): 273–74; 4 (1938): 66, 71; 4 (1939): 284; 5 (1939): 161; 5 (1940): 233, 247; 5 (1940): 243–44; 6 (1940): 85, 87; 6 (1940): 176; 6 (1941): 282–83; 6 (1941): 57; 7 (1941): 79; 7 (1941): 184–85.

56. WPA, Division of Information, Box 56, File 780-A, *Amarillo Sunday News & Globe*, 5 February 1939, 1.

57. "WPA Index," Texas; and WPA Monthly Reports, Texas: #20,449 and #30,976, January, 1939; June, 1939; September 1939; December, 1939; March, 1940; June, 1940; September, 1940; October, 1940; February, 1941; and #50,416, March, 1941; July, 1941.

58. "WPA Index," Texas; and WPA Monthly Reports: Texas, #30,067 and #40,343, September, 1938; December, 1938; March, 1939; December, 1939; March, 1940; June, 1940; September, 1940; December, 1940.

59. "WPA Index," Texas; and WPA Monthly Reports: Texas, #13,543, November, 1939; December, 1939; March, 1940; June, 1940; September, 1940; March, 1941; #75,021, September, 1938; December, 1938; March, 1939; June, 1939.

60. "WPA Index," Texas; WPA Monthly Reports: Texas, #20,319, August, 1939; #30,562, March, 1939; June, 1939; September, 1939; December, 1939; April, 1940; #40,670, June, 1940; September, 1940; December 1940; March, 1941.

61. "WPA Index," Texas; WPA Monthly Reports: Texas, #20,718, January, 1939; #30,625, December, 1939; March, 1940; June, 1940; September, 1940; December, 1940; March, 1941; "American Prehistory," 32, 41–43; NA, WPA, Division of Women and Professional, Box 57, File #14176, #13740, #6043; and *American Antiquity* 1 (1935): 134–35; 6 (1940): 84; 6 (1941): 274–75; 6 (April 1941): 351, 356–57.

62. *American Antiquity* 2 (1936): 52; "American Prehistory," 34–41; "Archaeological Projects Seek Light," 3; NA, WPA, Division of Information, Box 56, 780-A; "WPA Index," Oklahoma; and WPA Monthly Reports: Oklahoma, #50,291, July, 1935; December, 1938; March, 1939; September, 1939; December, 1939; March, 1940; June, 1940; September, 1940; December, 1940; March, 1941; June, 1941; September 1941.

63. WPA, Division of Information, Box 56, File 780-A, Oklahoma.

64. "WPA Index," Oklahoma; WPA Monthly Reports: Oklahoma, #20,705, #30,179, September, 1938; June, 1940; and #40,705, September, 1940; December, 1940; March, 1941; June, 1941; August, 1941; November, 1941; February, 1942; March, 1942; and WPA Division of Information, Box 56, File 780-A, Oklahoma.

65. *American Antiquity* 3 (1937): 83–84, 193; 5 (1939): 27–30.

66. "WPA Index"; WPA Monthly Reports: New Jersey, #12,282 and #30,579, December, 1936; May, 1937; November, 1937; February, 1938; June, 1938; September, 1938; December, 1938; March, 1939; September, 1939; December, 1939; March, 1940; June, 1940; September, 1940; December, 1940; and March, 1941; Dorothy Cross, "The Effect of the Abbott Farm on Eastern Chronology," *Proceedings of the American Philosophical Society* 86 (1943): 315–19; Nathaniel Knowles, "Cultural Stratification on the Trenton Bluff," *American Anthropologist* 43 (1941): 610–16; "American Prehistory," 50–53; WPA, Division of Women and Professional, Box 57, File #13612; *American Antiquity* 1 (1935): 120; 2 (1937): 227; 3 (1938): 280; 4 (1939): 103–104; 5 (1939): 158; 6 (1940): 81; and WPA, Division of Information, Box 56, File 780-A, New Jersey.

67. *American Antiquity* 1 (1935): 137–38; 2 (1936): 54; 2 (1937): 226.

68. *American Antiquity* 1 (1935): 58–59; 3 (1937): 85; 3 (1938): 279; 4 (1939): 357–58; 5 (1939): 157; 5 (1940): 205, 234; 5 (1940): 340; 6 (1940): 81; 6 (1941): 274; 7 (1941): 75–76.

69. *American Antiquity* 1 (1935): 125; 1 (1936): 324; 2 (1936): 53–55; 2 (1937): 226; 2 (1937): 311–12; 3 (1938): 279–80; 4 (1939): 287; 6 (1940): 174–75; 6 (1941): 272–73.

70. *American Antiquity* 1 (1935): 127–28; 5 (1939): 164–65; 5 (1940): 238–39; 6 (1940): 82; 6 (1941): 276; 7 (1941): 77; 7 (1941): 181–82.

71. *American Antiquity* 5 (1939): 164–65; 5 (1940): 238–39; and WPA, Division of Information, Box 56, File 780-A.

72. "WPA Index"; WPA Monthly Reports: Missouri, #40,077, #50,129, December, 1939; January, 1940; February, 1940; March, 1940; April, 1940; May, 1940; June, 1940; July, 1940; August, 1940; September, 1940; October, 1940; November, 1940; December, 1940; January, 1941; February, 1941; March, 1941; April, 1941; May, 1941; June, 1941; July, 1941; August, 1941; September, 1941; October, 1941; November, 1941; December, 1941; January, 1942; February, 1942; March, 1942; April, 1942; and *American Antiquity* 6 (1940): 63–64; 7 (1941): 77–78; 7 (1941): 411.

73. For one clear example, see "WPA Index"; WPA Monthly Reports: Missouri, December, 1940.

74. *American Antiquity* 1 (1935): 116; 1 (1936): 324.

75. *American Antiquity* 6 (1941): 279.

76. *American Antiquity* 4 (1949): 283–84.

77. *American Antiquity* 6 (1941): 279.

78. "American Prehistory," 63–64; and *American Antiquity* 1 (1935): 127; 2 (1936): 60; 4 (1939): 282–83; 6 (1941): 354.

79. "WPA Index"; WPA Monthly Reports: South Dakota, #30,506, June, 1939; September, 1939; December, 1939; June, 1940; Division of Women and Professional, Box 57, File #13613; and "American Prehistory," 63–64.

80. "WPA Index"; WPA Monthly Reports: South Dakota, #30,506, June 30, 1939; September, 1939; and *American Antiquity* 1 (1936): 130; 2 (1936): 59–60; 5 (1940): 245–46; 6 (1941): 354. Archaeology and anthropology at South Dakota at that time was, to quote Dr. Larry J. Zimmerman, Professor of Anthropology at the University of South Dakota, "very unorganized." (letter Zimmerman to author, 22 November 1985).

81. "Archaeological Projects Seek Light," 7; *American Antiquity* 1 (1936): 114–15; 2 (1936): 60; 3 (1938): 278; 4 (1939): 285–86; 5 (1939): 241–42; 6 (1940): 83; 6 (1941): 231–49; 7 (1941): 78; 7 (1941): 180.

82. "WPA Index"; WPA Monthly Reports: Minnesota, #75,063, October, 1939; November, 1940.

83. Ibid., #50,502, August, 1941; October, 1941. Wilford was a Harvard Ph.D. who served in a department of three to four persons. Particular data were courteously supplied by Professor Elden Johnson (letter Johnson to author, 18 November 1985).

84. "WPA Index"; WPA Monthly Reports: Iowa, #20,076, September, 1938; "Archaeological Projects Seek Light," 2; "American Prehistory," 60; and *American Antiquity* 1 (1935): 65; 2 (1936): 223–24; 5 (1939): 241.

85. *American Antiquity* 2 (1937): 310–11; 3 (1937): 82–83; 4 (1938): 73; 4 (1939): 283–84; 5 (1939): 68; 5 (1940): 244; 6 (1941): 280.

86. *American Antiquity* 3 (1937): 82–83.

87. *American Antiquity* 1 (1935): 115–16; 6 (1941): 274.
88. "WPA Index"; WPA Monthly Reports: Mississippi, #41,327, November, 1940; February, 941; May, 1941.
89. *American Antiquity* 6 (1940): 126; 2 (1936): 148.
90. *American Antiquity* 1 (1935): 175; 7 (1942): 232–54.
91. "WPA Index"; and WPA Monthly Reports: North Carolina, #40,074, September, 1940; "Turning Lost Pages of History," 2; and Division of Information, Box 56, 780-A.
92. "WPA Index"; WPA Monthly Reports: Arkansas, #30,277, October, 1939; November, 1939; December, 1939; January, 1940; "American Prehistory," 63; and *American Antiquity* 1 (1935): 52.
93. *American Antiquity* 6 (1940): 133–47.
94. *American Antiquity* 1 (1936): 197–214.
95. "Archaeological Projects Seek Light," 3.
96. "WPA Index"; WPA Monthly Reports: Michigan, #20,990, #40,034, #40,423, September, 1938; October 1938; November, 1938; December, 1938; January, 1939; February, 1939; March, 1939; April, 1939; May, 1939; June, 1939; July, 1939; August, 1939; September, 1939; October, 1939; November, 1939; December, 1939; January, 1940; February, 1940; March, 1940; April, 1940; May, 1940; June, 1940; July, 1940; August, 1940; September, 1940; October, 1940; November, 1940; December, 1940; January, 1941; February, 1941; March, 1941; April, 1941; May, 1941; June, 1941; and *American Antiquity* 1 (1935): 113–14.
97. NA, WPA, Division of Information, Box 56, File 780-A; and Division of Women and Professional, Box 57, File #13612.
98. NA, WPA, Division of Information, Box 56, File 780-A; and Division of Women and Professional, Box 57, File #13612.
99. NA, WPA, Division of Information, Box 56, File 780-A; and Division of Women and Professional, Box 57, File #13612.
100. Denney, "Centennial"; and interview with Marvin Kivett.

Bibliography

This study draws upon materials from a variety of sources: archival, personal correspondence, primary and secondary accounts, and oral interviews. Many agencies in Washington, D.C., played a major role in this endeavor. The National Archives houses all of the relief program papers, an area known extensively by its then resident-expert Robert Kvasnicka. His expertise aided significantly my inquiries into FERA, CWA, WPA, CCC, NYA, and NPS holdings. The National Research Council acted in a kind and courteous manner. A special thanks goes to the Smithsonian Institution. The Institution's archives provided material traced only with great difficulty by Harry Heis. The National Anthropological Archives constitute a rich mine of information, including the Papers of Frank M. Setzler, Frank H. H Roberts, James A. Ford, Matthew W. Stirling, W. Duncan Strong, Frederick Johnson, and records of the Bureau of American Ethnology and the U.S. National Museum. The Franklin D. Roosevelt Library staff helped in the task of foraging through the pertinent papers of Harry Hopkins, Lorena Hickock, and John M. Carmody. The helpful, professional people who staff the many university libraries in this country cheerfully fulfilled endless requests and demands, either in person or through the mail. My gratitude and thanks goes out to all: the Columbia University Oral History Collection for the transcripts of interviews with Florence Kerr and John Carmody; the University of Kentucky for the William S. Webb Papers; the Department of Anthropology at the University of Washington; the Department of Anthropology at Ohio State University, Columbus; the University of Michigan Museums, Department of Anthropology, and Special Collections; Harvard University and the Peabody Museum; the Macon Chamber of Commerce; Ocmulgee National Monument; the Department of Anthropology at the University of Alabama; the Department of Anthropology at the University of Mississippi; the Department of Anthropology at the University of Georgia; the Department of Anthropology at the University of Minnesota; the Department of Anthropology at the University of Oregon; the Department of Anthropology at the University of South Dakota; the Department of Anthropology at the University of New Mexico; the staff of the National Park Service at Albuquerque; the staff of the Laboratory of Anthropology at the Museum of New Mexico; the Southeastern Archaeological Center on the Florida State University campus at Tallahassee, the repository for the papers of the Society for Georgia Archaeology; the Department of Anthropology at the University of North Carolina, Chapel Hill; the Department of Anthropology at the University of Chicago; the Department of Anthropology at the University of Missouri, Columbia; the Department of Anthropology at the University of Kansas; the

Department of Anthropology at the University of Arizona; the Department of Anthropology at the University of Nebraska, Lincoln; the Nebraska State Historical Society, home to the A. T. Hill and the John L. Champe Papers; the library at Grinnel College; the Arthur and Elizabeth Schlesinger Library at Radcliffe College, which houses the Ellen S. Woodward Papers; the Glenn A. Black Laboratory of Archaeology at Indiana University; The University Museum of Archaeology/Anthropology and the Department of Anthropology at the University of Pennsylvania; the Department of Anthropology at Texas Tech University; the staff of the St. Louis Science Center; the library of the University of California, Riverside; the delightful staff of the Bancroft Library at the University of California, Berkeley; the Central Library at the University of California, San Diego; the Museum of Geoscience at Louisiana State University and its director Robert W. Neumann; and the Oral History Research Office at the University of Memphis. Warmest regards and thanks go to the gentlemen of the archaeological community who shared their time, background, and papers. For them, I hold the utmost admiration. Gordon R. Willey took off from a busy schedule at Harvard. Frederick Johnson saw me at the most difficult time in Boston. Carlyle Smith shared his home and made me welcome. The late Marvin Kivett fit me into an incredibly busy schedule and provided the most astute insights about the Depression. I still remember that rainy night in Ann Arbor and the professional courtesy and kindness of James B. Griffin. He treated this junior historian with the same dignity that makes him a respected leader in the field. Vincenzo Petrullo patiently answered questions about the past. I cannot begin to thank the late Emil Haury for what he did for me. For many years, this giant in American archaeology patiently guided and mentored my every effort. George Quimby, Frederick Hulse, and Richard MacNeish have corresponded freely. Stephen Williams of the Peabody Museum at Harvard University has graciously supplied me with further information. A special thanks goes to Jesse Jennings, who has several times shared his home and time with me. I owe a debt of gratitude to all and thank you deeply. Your recollections made this chapter in the history of archaeology as vivid and comprehensible for me today as it was for you over fifty years ago. The following bibliography is subdivided into basic categories, listing a few primary government sources in addition to the more substantial repositories already identified. Specific documents or reports not appearing here are cited individually in the footnotes. Books and articles comprise the remainder of this listing.

PRIMARY SOURCES
Government Documents

National Planning Board. *Final Report: 1933–34.* Washington, DC: Government Printing Office, 1934.

National Research Council. "Conference on Southern Pre-history Held under the Auspices of the Division of Anthropology and Psychology, Committee on State Archaeology Surveys, National Research Council, Birmingham, Alabama, December 18, 19, and 20, 1932." Washington, DC.

———. "The Indianapolis Archaeological Conference Held under the Auspices of the Division of Anthropology and Psychology, Committee on State Archaeological Sur-

veys, National Research Council, Indianapolis, Indiana, December 6, 7, and 8, 1935." Washington, DC.
National Resources Planning Board. *The Economic Effects of the Federal Public Works Expenditures: 1933–1938.* Washington, DC: Government Printing Office, 1940.
Public Works Administration. *A Four Year Record of the Construction of Permanent and Useful Public Works.* Washington, DC: Government Printing Office, 1937.
Works Progress Administration. *Operation Procedure No. 4.* Washington, DC: Government Printing Office, 1936.
———. *Operation Procedure No. 4.* Washington, DC: Government Printing Office, 1938.

SECONDARY SOURCES
Books and Pamphlets

Adams, Henry H. *Harry Hopkins.* New York: G. P. Putnam's Sons, 1977.
Albright, Horace M. *The Birth of the National Park Service: The Founding Years, 1913–33.* Salt Lake City: House Bros., 1985.
"American Museum of Natural History." In *The New Columbia Encyclopedia* (1975).
Bacon, Francis. *The Great Instauration and New Atlantis.* Edited by J. Weinberger. Arlington Heights, IL: AMH Pub. Corp., 1980.
Barber, Bernard, and Walter Hirsch, eds. *The Sociology of Science.* New York: Free Press of Glencoe, 1962.
Barnes, Barry. *Scientific Knowledge and Sociological Theory.* London: Routledge and Kegan Paul, 1974.
———. *Sociology of Science.* Harmondsworth, Eng.: Penguin Books, 1972.
———. *T. S. Kuhn and Social Science.* New York: Columbia University Press, 1982.
Bates, Ralph S. *Scientific Societies in the United States.* 3rd ed. Cambridge: MIT Press, 1965.
Bell, Earl. *Chapters in Nebraska Archaeology.* Omaha: University of Nebraska Press, 1936.
Beider, Robert E., and Thomas B. Tax."From Ethnologists to Anthropologists: A Brief History of the American Ethnological Society." In John V. Murra, ed., *American Anthropology: The Early Years.* New York: West Pub. Co., 1976.
Bennett, John W., *Archaeological Explorations in Jo Daviess County, Illinois: The Work of William Baker Nickerson (1895–1902) and the University of Chicago (1926–32).* Chicago: University of Chicago Press, 1945.
Bernstein, Irving. *A Caring Society: The New Deal, the Worker, and the Great Depression.* Boston: Houghton Mifflin, 1985.
Berelson, Bernard. *Graduate Education in the United States.* New York: McGraw-Hill Book Co., Inc., 1962.
Biles, Roger. *Memphis In The Great Depression.* Knoxville: University of Tennessee Press, 1986.
Blakey, George T. *Hard Times and the New Deal in Kentucky, 1929–1939.* Lexington: University of Kentucky Press, 1986.
Boas, Franz. "Anthropology," *Encyclopaedia of the Social Sciences* 1 (1932).
Boas, Franz. "The History of Anthropology." In Regna Darnell, ed., *Readings In The History of Anthropology.* New York: Harper & Row, Pubs., 1974.

Braeman, John; Robert H. Bremner; and David Brody, eds. *Change and Continuity in Twentieth Century America: The 1920's.* Columbus: The Ohio State University Press, 1968.
———. *The New Deal: Volume One, the National Level.* Columbus: The Ohio State University Press, 1975.
Brew, J. O. "Introduction." In J. O. Brew, ed., *One Hundred Years of Anthropology.* Cambridge: Harvard University Press, 1968.
———. *One Hundred Years of Anthropology.* Cambridge: Harvard University Press, 1968.
Brinton, Daniel B. "American Languages and Why We Should Study Them." In Regna Darnell, ed., *Readings in the History of Anthropology.* New York: Harper & Row, Pubs., 1974.
Brose, David S. "The Northeastern United States." In James E. Fitting, ed., *The Development of North American Archaeology: Essays In The History Of Regional Tradition.* Garden City, NJ: Anchor Books, 1973.
Brothwell, Don, and Eric Higgs, eds. *Science in Archaeology: A Comprehensive Survey of Progress and Research.* 2nd ed., enl. New York: Praeger Books, 1970.
Bruce, Robert V. *The Launching of Modern American Science, 1846–1876.* Ithaca: Cornell University Press, 1987.
Caldwell, Joseph R., and Catherine McCann. *Irene Mound Site, Chatham County, Georgia.* Athens: University of Georgia Press, 1941.
Campbell, Christina M. *The Farm Bureau and the New Deal: A Study of the Making of National Farm Policy, 1933–40.* Urbana: University of Illinois Press, 1962.
Chafe, William Henry. *The American Women: Her Changing Social, Economic, and Political Roles, 1920–1970.* New York: Oxford University Press, 1972.
Chauvenet, Beatrice. *Hewett and Friends: A Biography of Santa Fe's Vibrant Era.* Santa Fe: Museum of New Mexico Press, 1983.
Christenson, Andrew L. *Tracing Archaeology's Past: The Historiography of Archaeology.* Carbondale: Southern Illinois University Press, 1989.
Clawson, Marion. *New Deal Planning: The National Resources Planning Board.* Baltimore: The Johns Hopkins University Press, 1981.
Cohen, Lizabeth. *Making a New Deal: Industrial Workers in Chicago, 1919–1939.* Cambridge: Cambridge University Press, 1990.
Cole, Fay-Cooper, and Thorne Deuel. *Rediscovering Illinois: Archaeological Explorations in and around Fulton County.* Chicago: University of Chicago Press, 1937.
Cole, John R. "Nineteenth Century Fieldwork, Archaeology, and Museum Studies: Their Role in the Four-Field Definition of American Anthropology." In John V. Murra, ed., *American Anthropology: The Early Years.* New York: West Pub. Co., 1976.
Committee for the Recovery of Archaeological Remains. "The Inter-Agency Archaeological Salvage Program." Columbia: University of Missouri Press, 1958.
Crane, Diana. *Invisible Colleges: Diffusion of Knowledge in Scientific Communities.* Chicago: University of Chicago Press, 1972.
Daniel, Glyn. "One Hundred Years of Old World Prehistory." In J. O. Brew, ed., *One Hundred Years of Anthropology.* Cambridge: Harvard University Press, 1968.
Darnell, Regna. "Daniel Brinton and the Professionalization of American Anthropology." In John V. Murra, ed., *American Anthropology: The Early Years.* New York: West Pub. Co., 1976.
———, ed. *Readings In The History Of Anthropology.* New York: Harper & Row, Pubs., 1974.

DeJarnette, David L. "Alabama Archaeology: A Summary." In James B. Griffin, ed., *Archaeology of the Eastern United States*. Chicago: University of Chicago Press, 1952.
DeJarnette, David L., and Steve B. Wimberly. "The Bessemer Site: Excavation of Three Mounds and Surrounding Village Areas near Bessemer, Alabama." *Geological Survey of Alabama* (1941).
De Laguna, Frederica, ed. *Selected Papers from the American Anthropologist: 1888-1920*. Evanston, IL: Row, Peterson and Co., 1960.
De Reuck, Anthony, and Julie Knight, eds. *Communication in Science: Documentation and Automation*. New York: Little, Brown & Co., 1967.
Derber, Milton. "The New Deal and Labor." In John Braemen, Robert H. Bremner, and David Brody, eds., *The New Deal: Volume One, The National Level*. Columbus: The Ohio State University Press, 1975.
"The Development of Anthropology." In Frederica De Laguna, ed., *Selected Papers from the American Anthropologist, 1888-1920*. Evanston, IL: Row, Peterson and Co., 1960.
Devan, William C. *Higher Education in Twentieth Century America*. Cambridge: Harvard University Press, 1965.
Dupree, A. Hunter. *Science in the Federal Government: A History Of Policies And Activities To 1940*. Cambridge: Harvard University Press, 1957.
Edge, David O., and Michael J. Mulkay. *Astronomy Transformed: The Emergence of Radio Astronomy in Britain*. New York: John Wiley & Sons, 1956.
Ekirch, Arthur A., Jr. *Ideologies and Utopia: The Impact of the New Deal on American Thought*. Chicago: Quadrangle Books, 1969.
Fitting, James E., ed. *The Development of North American Archaeology: Essays In The History Of Regional Traditions*. Garden City, NJ: Anchor Books, 1973.
Fowke, Gerard. *Archaeological History of Ohio*. Columbus: The Ohio State University Press, 1902.
Fraser, Steve, and Gary Gerstle. *The Rise and Fall of the New Deal Order, 1930-1980*. Princeton: Princeton University Press, 1989.
Frison, George C. "The Plains." In James E. Fitting, ed., *The Development of North American Archaeology: Essays In The History Of Regional Traditions*. Garden City, NJ: Anchor Books, 1973.
Galbraith, John K. *The Great Crash*. 3rd ed. Boston: Houghton Mifflin Co., 1972.
Garraty, John A. *The Great Commonwealth: 1877-1890*. New York: Harper & Row, Pubs., 1968.
―――. *Unemployment in History: Economic Thought and Public Policy*. New York: Harper & Row, Pubs., 1978.
―――. *The Great Depression: An Inquiry into the causes, course, and consequences of the Worldwide Depression of the Nineteen-Thirties, as seen by contemporaries and in the light of History*. San Diego: Harcourt Brace Jovanovich, Pubs., 1986.
Givens, Douglas R. *Alfred Vincent Kidder and the Development of Americanist Archaeology*. Albuquerque: University of New Mexico Press, 1992.
Gladwin, Harold S., and Winifred Gladwin. *A Method for the Designation of Cultures and Their Variations*. Pasadena: Medallion Papers, 1928.
Goddard, Pliny Earle. "The Present Condition of Our Knowledge of North American Languages." In Frederica De Laguna, ed., *Selected Papers from the American Anthropologist: 1888-1920*. Evanston, IL: Row, Peterson and Co., 1960.

Goetzman, William H. *Army Exploration in the American West: 1803–1863*. New Haven: Yale University Press, 1959.

———. *Exploration and Empire: The Explorer and the Scientist in the Winning of the American West*. New York: Alfred A. Knopf, 1966.

Goldschmidt, Walter, ed. *The Uses of Anthropology*. Washington, DC: American Anthropological Association, 1979.

Graham, Otis L., and Meghan Robinson Wander, eds. *Franklin D. Roosevelt: His Life and Times, an Encyclopedic View*. Boston: G. K. Hall & Co., 1985.

Grantham, Dewey W., Jr. "The Twentieth Century South." In Arthur S. Link and Patrick W. Rembert, eds., *Writing Southern History: Essays in Historiography in Honor of Fletcher M. Green*. Baton Rouge: Louisiana State University Press, 1965.

Grayson, D. K. *The Establishment of Human Antiquity*. New York: Academic Press, 1983.

Griffin, James B., ed. *Archaeology of the Eastern United States*. Chicago: University of Chicago Press, 1952.

———. *The Fort Ancient Aspect: Its Cultural and Chronological Position in Mississippi Valley Archaeology*. Ann Arbor: University of Michigan Press, 1943.

Gruber, Jacob W. "Archaeology, History, and Culture." In David J. Meltzer, Don D. Fowler, and Jeremy A. Sabloff, eds., *American Archaeology Past and Future: A Celebration of the Society for American Archaeology, 1935–1985*. Washington, DC: Smithsonian Institution Press, 1986.

Guralnik, Stanley M. "The American Scientist in Higher Education, 1820–1910." In Nathan Reingold, ed. *The Sciences in the American Context: New Perspective*. Washington, DC: Smithsonian Institution Press, 1979.

Guthe, Carl. "Twenty-Five Years of Archaeology in the Eastern United States." In James B. Griffin, ed., *Archaeology of the Eastern United State*. Chicago: University of Chicago Press, 1952.

Hagstrom, Warren O. *The Scientific Community*. New York: Basic Books, 1943.

Hallowell, A. Irving. "The Beginnings of Anthropology in America." In Frederica De Laguna, ed., *Selected Papers from the American Anthropologist: 1888–1920*. Evanston, IL: Row, Peterson and Co., 1960.

Hally, David J., ed. *Ocmulgee Archaeology: 1936–1986*. Athens: University of Georgia Press, 1994.

Hamerow, Theodore S. *Reflections on History and Historians*. Madison: University of Wisconsin Press, 1987.

Haskell, Thomas L. *The Emergence of Professional Social Science: The American Social Science Association and the Nineteenth-Century Crisis of Authority*. Chicago: University of Chicago Press, 1977.

Hauptman, Lawrence M. *The Iroquois and the New Deal*. Syracuse: Syracuse University Press, 1977.

Hawley, Ellis W. "The New Deal and Business." In John Braemen, Robert H. Bremner, and David Brody, eds., *The New Deal: Volume One, The National Leve*. Columbus: The Ohio State University Press, 1975.

Hinsley, Curtis M., Jr. "Edgar Lee Hewett and the School of American Archaeology in Santa Fe, 1906–1912." In David J. Meltzer, Don D. Fowler, and Jeremy A. Sabloff, eds., *American Archaeology Past and Future: A Celebration of the Society for American Archaeology, 1935–1985*. Washington, DC: Smithsonian Institution Press, 1986.

———. "Revising and Revisioning the History of Archaeology: Reflections on Region and Context." In Andrew L. Christenson, ed., *Tracing Archaeology's Past: The Historiography of Archaeology*. Carbondale: Southern Illinois University Press, 1989.

———. *Savages and Scientists: The Smithsonian Institution and the Development of American Anthropology, 1846–1910*. Washington, DC: Smithsonian Institution Press, 1981.

Hodges, James A. *New Deal Labor Policy and the Southern Cotton Textile Industry: 1933–1941*. Knoxville: University of Tennessee Press, 1986.

Holmes, Michael S. *New Deal in Georgia: An Administrative History*. Westport, CT: Greenwood Press, 1975.

Holmes, W. H. "The World's Fair Congress of Anthropology." In Frederica De Laguna, ed., *Selected Papers from the American Anthropologist: 1888–1920*. Evanston, IL: Row, Peterson and Co., 1960.

Holt, James. "The New Deal and the American Anti-Statist Tradition." In John Braeman, Robert H. Bremner, and David Brody, eds., *The New Deal: Volume One, The National Leve*. Columbus: The Ohio State University Press, 1975.

Hopkins, Harry L. *Spending to Save: The Complete Story of Relief*. New York: W. W. Norton & Co., 1936.

Hosmer, Charles B., Jr. *Preservation Comes of Age: A History of the Preservation Movement before Williamsburg*. New York: Putnam Books, 1965.

———. *Preservation Comes of Age: From Williamsburg to the National Trust, 1926–1949*. Charlottesville: University of Virginia Press, 1981.

Jaher, Frederic Cople, ed. *The Rich, the Well-born, and the Powerful: Elites and Upper Classes In History*. Urbana: University of Illinois Press, 1973.

Jennings, Jesse D. *Accidental Archaeologist: Memoirs of Jess D. Jennings*. Salt Lake City: University of Utah Press, 1994.

———. "American Archaeology, 1930–1985." In David J. Meltzer, Don D. Fowler, and Jeremy A. Sabloff, eds., *American Archaeology Past and Future: A Celebration of the Society for American Archaeology, 1935–1985*. Washington, DC: Smithsonian Institution Press, 1986.

———. "Prehistory of the Lower Mississippi Valley." In James B. Griffin, ed., *Archaeology of the Eastern United States*. Chicago: University of Chicago Press, 1952.

Jennings, Jesse; Elden Johnson; Arden R. King; Robert F. Spencer; Theodore Stern; Kenneth M. Stewart; and William J. Wallace. *The Native Americans: Ethnology and Background of the North American Indians*. 2nd ed. New York: Harper & Row, Pubs., 1977.

Johnson, Jay K., ed. *The Development of Southeastern Archaeology*. Tuscaloosa: The University of Alabama Press, 1993.

Judd, Neil. *The Bureau of American Ethnology: A Partial History*. Norman: University of Oklahoma Press, 1968.

Kardiner, Abram, and Edward Preble. *The Studied Man*. New York: Mentor Books, 1961.

Kehoe, Alice B. "The Paradigmatic Vision of Archaeology: Archaeology as a Bourgeois Science." In Jonathan E. Reyman, ed., *Rediscovering Our Past: Essays on the History of American Archaeology*. Newcastle, Eng.: Athenaeum Press, 1992.

Kelley, Jane H., and Marsha P. Hanen. *Archaeology and the Methodology of Science*. Albuquerque: University of New Mexico Press, 1988.

Kevles, Daniel J. *The Physicists: The History of a Scientific Community in Modern America*. New York: Alfred A. Knopf, 1978.

Kidder, Alfred V. *An Introduction to the Study of Southwestern Archaeology.* New Haven: Yale University Press, 1924.

Kirkendall, Richard S. "The New Deal and Agriculture." In John Braemen, Robert H. Bremner, and David Brody, eds., *The New Deal: Volume One, The National Level.* Columbus: The Ohio State University Press, 1975.

Kneburg, Madeline. "The Tennessee Area." In James B. Griffin, ed., *Archaeology of the Eastern United States.* Chicago: University of Chicago Press, 1952.

Kohler, Robert. *From Medical Chemistry to Biochemistry.* London: Cambridge University Press, 1982.

Kohlstedt, Sally G. "A Step Toward Scientific Self-Identity in the United States: The Failure of the National Institute, 1844." In Nathan Reingold, ed., *Science in America Since .* New York: Science History Pubs., 1976.

Kohlstedt, Sally Gregory, ed. *The Origins of Natural Science in America.* Washington, DC: Smithsonian Institution Press, 1991.

Kroeber, A. L. "A History of Personality of Anthropology." In Regna Darnell, ed., *Readings In The History Of Anthropology.* New York: Harper & Row, Pubs., 1974.

Krueger, Thomas A., and William Glidden. "The New Deal Intellectual Elite: A Collective Portrait." In Frederic Cole Jaher, ed., *The Rich, the Well-born, and the Powerful: Elites and Upper Classes in History.* Urbana: University of Chicago Press, 1973.

Kuhn, Thomas S. *The Essential Tension: Selected Studies in Scientific Tradition and Change.* Chicago: University of Chicago Press, 1977.

———. "Reflections on My Critics." In Imre Lakatos and Alan Musgrave, eds., *Criticism and the Growth of Knowledge: Proceedings of the International Colloquium in the Philosophy of Science, London, 1965, Vol. 4.* Cambridge: Cambridge University Press, 1970.

———. *The Structure of Scientific Revolutions.* 2nd ed., enl. Chicago: University of Chicago Press, 1970.

Kurzman, Paul H. *Harry Hopkins and the New Deal.* Fairlawn, NJ: R. E. Burdick, Inc., 1974.

Lakatos, Imre, and Alan Musgrave, eds. *Criticism and the Growth of Knowledge: Proceedings of the International Colloquium in the Philosophy of Science, London, 1965, Vol. 4.* Cambridge: Cambridge University Press, 1970.

Lee, Ronald. *United States: Historical and Archaeological Monuments.* Mexico, DF: Instituto Panamericano De Geografia E Historica, 1975.

Lesser, Alexander. "The American Ethnological Society: The Columbia Phase, 1906–1946." In John V. Murra, ed., *American Anthropology: The Early Years.* New York: West Pub. Co., 1976.

Leuchtenberg, William E. *Franklin D. Roosevelt and the New Deal: 1932–1940.* New York: Harper & Row, Pubs., 1963.

———. *The Perils of Prosperity: 1914–1932.* Chicago: University of Chicago Press, 1958.

Levine, David O. *The American College and the Culture of Aspiration, 1915–1940.* Ithaca: Cornell University Press, 1986.

Lewis, T. M. N., and Madeline D. Kneberg Lewis. *The Prehistory of the Chickamagua Basin in Tennessee.* 2 Vols. Edited by Lynne P. Sullivan. Knoxville: University of Tennessee Press, 1995.

Louchheim, Katie, ed. *The Making of the New Deal: The Insiders Speak.* Cambridge: Harvard University Press, 1983.

MacMahon, Arthur W.; John D. Millet; and Gladys Ogden. *The Administration of Federal Work Relief.* New York: Da Capo Press, 1971.
MacNeish, Richard S. *The Science of Archaeology.* Belmont, CA: Duxbury Press, 1978.
Madge, John. *Origins of Scientific Sociology.* New York: Free Press of Glencoe, 1962.
Martin, Paul S.; Donald Collier; and George I. Quimby, Jr. *Indians Before Columbus: Twenty Thousand Years of North American History Revealed by Archaeology.* Chicago: University of Chicago Press, 1947.
Mason, Otis T. "The Technic of Aboriginal American Basketry." In Frederica De Laguna, ed., *Selected Papers from the American Anthropologist: 1888-1920.* Evanston, IL: Row, Peterson and Co., 1960.
Masterman, Margaret. "The Nature of a Paradigm." In Irme Lakatos and Alan Musgrave, eds., *Criticism and the Growth of Knowledge.* Cambridge: Cambridge University Press, 1970.
McElvaine, Robert. *The Great Depression, 1929-1941.* New York: Times Books, 1984.
McJimsey, George. *Harry Hopkins: Ally of the Poor and Defender of Democracy.* Cambridge: Harvard University Press, 1987.
Meadows, A. J. *Communication in Science.* Stoneham, MA: Butterworth, 1974.
Meltzer, David J.; Don D. Fowler; and Jeremy Sabloff, eds. *American Archaeology Past and Future: A Celebration of the Society for American Archaeology, 1935-1985.* Washington, DC: Smithsonian Institution Press, 1986.
Mentzel, Herbert. *The Flow of Information Among Scientists.* 2 vols. New York: Columbia University Press, 1958.
———. "Planning the Consequences of Unplanned Action in Scientific Communication." In Anthony De Reuck and Julie Knight, eds., *Communication in Science: Documentation and Automation.* New York: Little, Brown & Co., 1967.
Midgette, Nancy Smith. *To Foster The Spirit Of Professionalism: Southern Scientists and State Academies of Science.* Tuscaloosa: University of Alabama Press, 1991.
Miller, Howard S. *Dollars for Research: Science and Its Patrons in Nineteenth Century America.* Seattle: University of Washington Press, 1970.
———. "Science and Private Agencies." In by D. Van Tassel and Michael G. Hall, eds., *Science and Society in the United States.* Homewood, IL: Dorsey Press, 1966.
Miller, Jimmy H. *The Life of Harold Sellers Colton.* Tsaile, AZ: Navajo Community College Press, 1991.
Minton, John Dean. *The New Deal in Tennessee: 1932-1938.* New York: Garland Pub., Inc., 1979.
Moore, Wilbert E. *The Professions: Roles and Rules.* New York: Russell Sage Foundation, 1970.
Murra, John V. "American Anthropology: The Early Years." In John V. Murra, ed., *American Anthropologist: The Early Years.* New York: West Pub. Co., 1976.
Murra, John V., ed. *American Anthropology: The Early Years.* New York: West Pub. Co., 1976.
"Obituary of John Wesley Powell." In Frederica De Laguna, ed., *Selected Papers from the American Anthropologist: 1888-192.* Evanston, IL: Row, Peterson and Co., 1960.
Nelson, C. E., and D. K. Pollock, eds. *Communication Among Scientists and Engineers.* Lexington, MA.: D. C. Heath, 1970.
Neumann, Robert W. *An Introduction to Louisiana Archaeology.* Baton Rouge: Louisiana State University Press, 1984.

Oleson, Alexander, and Sanborn C. Brown, eds. *The Pursuit Of Knowledge In The Early American Republic: American Scientific and Learned Societies from Colonial Times to the Civil War*. Baltimore: Johns Hopkins Press, 1976.

Parker, Arthur C. *The Archaeological History of New York*. Albany: New York State Museum Bulletin, 1922.

Parsons, Talcott. *Essays in Sociological Theory*. rev. ed. Glencoe, IL: Free Press, 1954.

———. "Professions." In *The International Encyclopedia of the Social Sciences*, Vol. 12 (1932).

Patterson, James T. *Congressional Conservatism and the New Deal: The Growth of the Conservative Coalition in Congress, 1933–1939*. Lexington: University of Kentucky Press, 1967.

———. *The New Deal and the States: Federalism in Transition*. Princeton: Princeton University Press, 1969.

Penniman, T. K. *A Hundred Years of Anthropology*. 3rd ed., rev. London: Gerald Duckworth & Co., 1965.

Polenberg, Richard. "The Decline of the New Deal." In John Braemen, Robert H. Bremner, and David Brody, eds., *The New Deal: Volume One, the National Level*. Columbus: The Ohio State University Press, 1975.

Pope, G. D., Jr. *Ocmulgee National Monument—Georgia; National Park Service Handbook Series #24*. Washington, DC: Government Printing Office, 1958.

Poppendieck, Janet. *Breadlines Knee-Deep in Wheat: Food Assistance in the Great Depression*. New Brunswick: Rutgers University Press, 1986.

Post, Robert. "Science, Public Policy, and Popular Precepts: Alexander Dallas Bache and Alfred Beach as Symbolic Adversaries." In Nathan Reingold, ed., *The Sciences in the American Context: New Perspectives*. Washington, DC: Smithsonian Institution Press, 1979.

Quimby, George I., Jr. "A Brief History of WPA Anthropology." In Walter Goldschmidt, ed., *The Uses of Anthropology*. Washington, DC: American Anthropological Association, 1979.

Reagan, Michael D. *Science and the Federal Patron*. New York: Oxford University Press, 1969.

Reingold, Nathan. "Alexander Dallas Bache." In *Dictionary of Scientific Biography*, Vol. 12 (1970).

———. "Definitions and Speculations: The Professionalization of Science in the Nineteenth Century." In Alexandra Oleson and Sanborn C. Brown, eds., *The Pursuit Of Knowledge In The Early American Republic: American Scientific and Learned Societies from Colonial Times to the Civil War*. Baltimore: Johns Hopkins University Press, 1976.

———. "Joseph Henry." In *Dictionary of Scientific Biography*, Vol. 6 (1970).

———, ed. *The Papers of Joseph Henry*. Vol. 1: *December 1797–October 1832: The Albany Years*. Washington, DC: Smithsonian Institution Press, 1972; Vol. 2: *November 1832–December 1835: The Princeton Years*. Washington, DC: Smithsonian Institution Press, 1975; Vol. 3: *January 1836–December 1837: The Princeton Years*. Washington, DC: Smithsonian Institution Press, 1979; Vol. 4: *January 1838–December 1840: The Princeton Years*. Washington, DC: Smithsonian Institution Press, 1981.

———. *Science in America Since 1820*. New York: Science History Pubs., 1976.

———. *The Sciences in the American Context: New Perspectives*. Washington, DC: Smithsonian Institution Press, 1979.

———. *Science in Nineteenth Century America, A Documentary History*. New York: Hill and Wang, 1964.

Reyman, Jonathan E., ed. *Rediscovering Our Past: Essays on the History of American Archaeology.* Newcastle, Eng.: Athenaeum Press, 1992.
Romasco, Albert U. *The Politics of Recovery: Roosevelt's New Deal.* New York: Oxford University Press, 1983.
Rossiter, Margaret W. *Women Scientists In America: Struggles And Strategies To 1940.* Baltimore: Johns Hopkins University Press, 1982.
Rowe, Chandler W. "Woodland Cultures of Eastern Tennessee." In James B. Griffin, ed., *Archaeology of the Eastern United States.* Chicago: University of Chicago Press, 1952.
Ruse, Michael. *The Darwinian Revolution: Science Red in Tooth and Claw.* Chicago: University of Chicago Press, 1979.
Salmond, John A. "Aubrey Williams: Atypical New Dealer?" In John Braeman, Robert H. Bremner, and David Brody, eds., *The New Deal: Volume One, The National Level.* Columbus: The Ohio State University Press, 1975.
———. *The Civilian Conservation Corps, 1933–1942: A New Deal Case Study.* Durham: Duke University Press, 1967.
———. *A Southern Rebel: The Life and Times of Aubrey Willis Williams, 1890–1965.* Chapel Hill: University of North Carolina Press, 1983.
Saloutos, Theodore. *The American Farmer and the New Deal.* Ames, IA: Iowa State University Press, 1982.
Schiffer, M. B., ed. *Advanced in Archaeological Method and Theory.* New York: Academic Press, 1983.
Schlesinger, Arthur M., Jr. *The Age of Roosevelt.* Vol. 1: *Crisis of the Old Order.* Boston: Houghton Mifflin Co., 1957.
———. *The Age of Roosevelt.* Vol. 2: *The Coming of the New Deal.* Boston: Houghton Mifflin Co., 1958.
Schwartz, Bonnie Fox. *The Civil Works Administration, 1933–1934.* Princeton: Princeton University Press, 1984.
Schwartz, Douglas W. *Conceptions of Kentucky Prehistory: A Case Study in the History of Archaeology.* Lexington: University of Kentucky Press, 1967.
Sherwood, Robert. *Roosevelt and Hopkins: An Intimate History.* Rev. Ed. New York: Harper & Row, Pubs., 1950.
Sprague, Roderick. "The Pacific Northwest." In James E. Fitting, ed., *The Development of North American Archaeology: Essays In The History Of Regional Traditions.* Garden City, NJ: Anchor Books, 1973.
Stanton, William. *The Leopard's Spots: Scientific Attitudes Toward Race in America, 1815–59.* Chicago: University of Chicago Press, 1960.
Stegner, Wallace. *Beyond the Hundredth Meridian: John Wesley Powell and the Second Opening of the West.* Boston: Houghton Mifflin, 1954.
———. "John Wesley Powell." In *Dictionary of Scientific Biography*, Vol. 11 (1970).
Sterud, E. L. "A Paradigmatic View of Prehistory." In Colin Renfrew, ed., *The Explanation of Culture Change.* London: Duckworth, 1973.
Stiebing, William H., Jr. *Uncovering the Past: A History of Archaeology.* Buffalo, NY: Prometheus Books, 1993.
Stirling, Matthew W. "Smithsonian Archaeological Projects Conducted under the Federal Emergency Relief Administration, 1933–34." In *Annual Report of the Board of Regents.* Washington, DC: Smithsonian Institution Press, 1934.

Stocking, George W., Jr. "Ideas and Institutions in American Anthropology: Thoughts Toward a History of the Interwar Years." In George W. Stocking, Jr., *Selected Papers from the American Anthropologist: 1921–1945* Washington, DC: American Anthropological Association, 1976.

———, ed. *Selected Papers from the American Anthropologist 1921–1945*. Washington, DC: American Anthropological Association, 1976.

———. *The Shaping of American Anthropology, 1883–1911: A Franz Boas Reader*. New York: Basic Books, 1974.

———. "Some Problems in the Understanding of Nineteenth Century Cultural Evolutionism." In Regna Darnell, ed., *Readings in the History of Anthropology*. New York: Harper & Row, Pubs., 1974.

Stoltman, James B. "The Southeastern United States." In James E. Fitting, ed., *The Development of North American Archaeology: Essays In The History Of Regional Traditions*. Garden City, NJ: Anchor Books, 1973.

Storr, Michael J. *The Beginning of Graduate Education in America*. Chicago: University of Chicago Press, 1953.

Tax, Sol. "The Setting of the Science of Man." In Sol Tax, ed., *Horizons of Change*. Chicago: Aldine Pub. Co., 1964.

Tax, Sol, ed. *Horizons of Anthropology*. Chicago: Aldine Pub. Co., 1964.

Taylor, Walter W. *A Study of Archaeology*. Carbondale, IL: Southern Illinois University Press, 1964.

Tobey, Ronald C. *The American Ideology of National Science, 1919–1930*. Pittsburgh: University of Pittsburgh Press, 1971.

———. *Saving the Prairies: The Life Cycle of the Founding School of American Plant Ecology, 1895–1955*. Berkeley: University of California Press, 1981.

Trigger, Bruce G. "History and contemporary American archaeology: a critical analysis." In C. C. Lamberg-Karlovsky and P. L. Kohl, eds., *Archaeological Thought in America*. Cambridge: Cambridge University Press, 1988.

———. *A History of Archaeological Thought*. Cambridge: Cambridge University Press, 1989.

Washburn, Wilcomb E. "Joseph Henry's Conception of the Purpose of the Smithsonian Institution." In Walter Muir Whitehill, ed., *A Cabinet of Curiosities*. Charlottesville: University of Virginia Press, 1967.

Van Tassel, D., and Michael G. Hall, eds. *Science and Society in the United States*. Homewood, IL: Dorsey Press, 1966.

Vesey, Lawrence R. *The Emergence of the American University*. Chicago: University of Chicago Press, 1965.

"Vincenzo Petrullo." In *American Men And Women Of Science* (1973).

Voget, Fred W. "Franz Boas." In *Dictionary of Scientific Biography*, Vol. 2 (1970).

Walker, Winslow W. "A Reconnaissance of Northern Louisiana Mounds." *Explorations and Field-Work of the Smithsonian Institution in 1931* (1931).

———. "Trailing the Moundbuilders of the Mississippi Valley." *Explorations and Field-Work of the Smithsonian Institution in 1932* (1932).

Ware, Susan. *Beyond Suffrage: Women in the New Deal*. Cambridge: Harvard University Press, 1981.

Washburn, Wilcomb E. *The Cosmos Club of Washington: A Centennial History, 1878–1978*. Washington, DC: Cosmos Club, 1978.

Wauchope, Robert. "Archaeological Survey of Northern Georgia." *Memoirs of the Society for American Archaeology*, Vol. 21 (1966).
Webb, W. S., and Charles E. Snow. *The Adena People with a Chapter on Adena Pottery and a Foreword to the New Edition by James B. Griffin*. Knoxville: University of Tennessee Press, 1974.
Webb, W. S., and Charles G. Wilder. *An Archaeological Survey of Guntersville Basin in the Tennessee River in Northern Alabama*. Lexington: University of Kentucky Press, 1951.
Whisenhunt, Donald W. *The Depression in the Southwest*. Port Washington, NY: National University Press, 1980.
Whiteford, Andrew H. "A Frame Of Reference For The Archaeology Of Eastern Tennessee." In James B. Griffin, ed., *Archaeology of the Eastern United States*. Chicago: University of Chicago Press, 1952.
Whitehill, Walter Muir, ed. *A Cabinet of Curiosities*. Charlottesville: University of Virginia Press, 1980.
Wiebe, Robert H. *The Search for Order, 1877–1920*. New York: Hill and Wang, 1967.
Willey, Gordon R. "One Hundred Years of American Archaeology." In J. O. Brew, ed., *One Hundred Years of Anthropology*. Cambridge: Harvard University Press, 1968.
———. *Portraits in American Archaeology: Remembrances of Some Distinguished Americanists*. Albuquerque: University of New Mexico Press, 1988.
Willey, Gordon R., and Jeremy A. Sabloff. *A History of American Archaeology*. 3rd ed., rev. New York: W. H. Freeman & Co., 1993.
Williams, Barbara. *Breakthrough: Women in Archaeology*. New York: Walker and Co., 1981.
Wimberly, Steve B., and Harry A. Tourtelot. "The McQuorquodale Mound: A Manifestation of the Hopewellian Phase in South Alabama." *Geological Survey of Alabama* (1941).
Wolters, Raymond. "The New Deal and the Negro." In John Braeman, Robert H. Bremner, and David Brody, eds., *The New Deal: Volume One, The National Level*. Columbus: The Ohio State University Press, 1975.
Woodbury, Richard B. *Alfred V. Kidder*. New York: Columbia University Press, 1973.
———. *Sixty Years of Southwestern Archaeology: A History of the Pecos Conference*. Albuquerque: University of New Mexico Press, 1993.
Woodward, C. Vann. *Origins of the New South: 1877–1913*. Baton Rouge: Louisiana State University Press, 1951.
Worster, Donald. *Dust Bowl: The Southern Plains in the 1930s*. New York: Oxford University Press, 1979.
Zinn, Howard, ed. *New Deal Thought*. Indianapolis: The Bobbs- Merrill Co., Inc., 1966.
Zochert, Donald. "Science and the Common Man in Ante-Bellum America." In Nathan Reingold, ed., *Science in America Since 1820*. New York: Science History Pubs., 1976.

Articles

Amick, Daniel J. "An Index of Scientific Elitism and the Scientist's Mission." *Social Studies of Science* 4 (1974): 1–16.
Ascher, Robert. "Archaeology and the Public Image." *American Antiquity* 25 (January 1960): 402–403.
Auerback, Lewis I. "Scientists in the New Deal: A Pre-War Episode in the Relations between Science and Government in the United States." *Minerva* 3 (1965): 457–82.

Augustine, Edgar E. "Recent Discoveries in Somerset County." *Pennsylvania Archaeologist* 8 (1938): 6–12.

Barber, Bernard. "Some Problems in the Sociology of the Professions." *Daedalus* 92 (1963): 669–88.

Beer, John J., and David W. Lewis. "Aspects of the Professionalization of Science." *Daedalus* 92 (1963): 764–84.

Ben-David, Joseph, and Awraham Zloczowee. "The Scientific Community in American and European Universities." *European Journal of Sociology* 3 (1960): 45–84.

Bender, Thomas. "Wholes and Parts: Continuing the Conversation." *Journal of American History* 74 (1987): 123–30.

Boulard, Garry. "'State of Emergency': Key West in the Great Depression." *Florida Historical Quarterly* 67 (1988): 166–83.

Bremer, William W. "Along the 'American Way': The New Deal's Work Relief Programs for the Unemployed." *Journal of American History* 62 (1975): 636–52.

Brew, J. O. "Salvage In River Basins: A World View." *Archaeology* 14 (1961): 233–35.

———. "Symposium On River Valley Archaeology." *American Antiquity* 12 (1947): 209–25.

Camfield, Thomas M. "The Professionalization of American Psychology, 1870–1917." *Journal of the History of the Behavioral Sciences* 9 (1973): 66–75.

Chaetlain, Verne. "The National Park Service and the New Deal." *OAH Newsletter* 13 (1985): 11–13.

Chapman, Carl H. "The Amateur Archaeological Society: A Missouri Example." *American Antiquity* 50 (1985): 241–48.

Coe, Michael D. "Matthew W. Stirling." *American Antiquity* 41 (1976): 67–73.

Collier, Donald, and Harry Tschopik, Jr. "The Role of Museums in American Anthropology." *American Anthropologist* 56 (1954): 768–79.

Cooper, Paul. "Report of Explorations." *Nebraska History Magazine* 20 (1939): 95–99.

Cross, Dorothy. "The Effect of the Abbott Farm on Eastern Chronology." *Proceedings of the American Philosophical Society* 86 (1943): 315–19.

Daniels, George H. "The Process of Professionalization in American Science: The Emergent Period, 1820–1860." *Isis* 58 (1967): 151–66.

Darnell, Regna. "The Emergence of Academic Anthropology at the University of Pennsylvania." *Journal of the History of the Behavioral Sciences* 6 (1970): 80–92.

———. "The Role of History of Anthropology in the Anthropology Curriculum." *Journal of the History of the Behavioral Sciences* 18 (1982): 265–71.

Denney, James. "History Observes a Centennial." *Magazine of the Midlands, Sunday Edition, The Omaha World-Herald* (24 September 1978): 19–22.

Dunnell, Robert C. "Archaeological Survey in the Lower Mississippi Alluvial Valley, 1940–1947: A Landmark Study in American Archaeology." *American Antiquity* 50 (1985): 297–300.

Edge, David. "Is there too much sociology of science?" *Isis* 74 (1983): 250–56.

Eggan, Fred. "Fay-Cooper Cole." *American Anthropology* 65 (1963): 641–48.

Evans, Clifford. "James Alfred Ford, 1911–1968." *American Anthropologist* 70 (1968): 1162.

Ezrahi, Yaron. "The Political Resources of American Science." *Science Studies* 1 (1971): 117–33.

Ford, James A. "Greenhouse: A Troyville–Coles Creek Period Site in Avoyelles Parish, Louisiana." *American Museum of Natural History Anthropological Papers* 44 (1951): 32.

Ford, James A., and George I. Quimby, Jr. "The Tchefuncte Culture: An Early Occupation of the Lower Mississippi Valley." *Memoirs of the Society for American Archaeology* 2 (1945).
Ford, James A., and Gordon R. Willey. "An Interpretation of the Prehistory of the Eastern United States." *American Anthropologist* 43 (1941): 325–63.
Freeman, John Finley. "Religion and Personality in the Anthropology of Henry Schoolcraft." *Journal of the History of the Behavioral Sciences* 1 (1965): 301–13.
Friedken, Noah E. "University Social Structure and Social Networks Among Scientists." *American Journal of Sociology* 83 (1978): 1445–65.
Gieryn, Thomas F., and Robert K. Merton. . "The Sociological Study of Scientific Specialties." *Social Studies of Science* 6 (1976): 445–70.
Gifford, Carol H., and Elizabeth A. Morris. "Digging for Credit: Early Archaeological Field Schools in the American Southwest." *American Antiquity* 50 (1985): 395–411.
Gladwohl, David M. "Marvin F. Kivett: 1917–1992." *American Antiquity* 59 (1994): 464–70.
Godoy, Ricardo. "Franz Boas and His Plans for an International School of American Archaeology and Ethnology in Mexico." *Journal of the History of the Behavioral Sciences* 13 (1977): 228–42.
Griffin, James B. "Carl Eugen Guthe." *American Antiquity* 41 (1976): 168–77.
———. "The Chronological Position and Ethnological Relationships of the Fort Ancient Aspect." *American Antiquity* 2 (1937): 273–76.
———. "Eastern North American Archaeology: A Summary." *Science* 156 (1967): 175–91.
———. "The Formation of the Society for American Archaeology." *American Antiquity* 50 (1985): 261–71.
———. "The Pursuit of Archaeology in the United States." *American Anthropologist* 61 (1959): 379–89.
Griffith, Belver C.; Marilyn J. Jahn;and A. James. Miller. "Informal Contacts in Science: A Probabalistic Model for Communication Processes." *Science* 173 (1971): 164–66.
Guthe, Carl. "The Basic Needs of American Archaeology." *Science* 90 (1939): 528–30.
———. "Reflections On The Founding Of The Society For American Archaeology." *American Antiquity* 32 (1967): 433–40.
Haag, William G. "Federal Aid to Archaeology in the Southeast, 1933–1942." *American Antiquity* 50 (1985): 272–80.
———. "Twenty Five Years of Eastern Archaeology." *American Antiquity* 27 (1961): 16–23.
———. "William Snyder Webb, 1882–1964." *American Antiquity* 30 (1965): 470–73.
Hagstrom, Warren O. "Competition in Science." *American Sociological Review* 39 (1974): 1–18.
Hallowell, A. Irving. "The History of Anthropology as an Anthropological Problem." *Journal of the History of the Behavioral Sciences* 1 (1965): 24–38.
Harrington, J. C. "Archaeology as an Auxiliary Science to American History." *American Anthropologist* 57 (1955): 1121–30.
Haury, Emil W. "Reflections: Fifty Years of Southwestern Archaeology." *American Antiquity* 50 (1985): 383–94.
Hill, A. T., and Paul Cooper. "The Archaeological Campaign of 1937 by the Nebraska State Historical Society." *Nebraska History Magazine* 18 (1937): 243–47.

———. "The Champe Site." *Nebraska History Magazine* 18 (1937): 223–25.
———. "The Schrader Site." *Nebraska History Magazine* 17 (1936): 223–25.
Hollinger, David. "T. S. Kuhn's Theory of Science and its Implications for History." *American Historical Review* 78 (1973): 370–93.
Hughes, Everett C. "Professions." *Daedalus* 92 (1963): 655–68.
Jennings, Jesse. "The Archaeological Survey of the Natchez Trace." *American Antiquity* 9 (1944): 408–14.
———. "River Basin Surveys: Origins, Operations, and Results, 1945–1969." *American Antiquity* 50 (1985): 281–96.
Johnson, Charles W. "The Army and the Civilian Conservation Corps, 1933–42." *Prologue* 4 (1972): 139–296.
Johnson, Frederick. "Archaeology in an Emergency." *Science* 17 (1966): 1592–97.
———. "The Work of the Committee for the Recovery of Archaeological Remains: Aims, History, and Activities to Date." *American Antiquity* 12 (1947): 212–15.
———. "The Inter-Agency Archaeological Salvage Program In The United States." *Archaeology* 4 (1951): 25–40.
———. "A Quarter Century of Growth in American Archaeology." *American Antiquity* 27 (1961): 1–6.
Johnson, Frederick; Emil W. Haury;, and James B. Griffin. "The Planning Committee of the Society for American Archaeology: A Preliminary Report." *American Antiquity* 10 (1945): 220.
———. "Report of the Planning Committee." *American Antiquity* 11(1945): 142–44.
Judd, Neil. "The Present Status of Archaeology in the United States." *American Anthropologist* 31 (1929): 401–407.
Kahler, Herbert E. "The Role of the National Park Service in River Basin Archaeology with Particular Reference to Inter-Bureau Agreements." *American Antiquity* 12 (1947): 215–16.
Kellar, James H. "Glenn A. Black." *Indiana Magazine of History* 63 (1967): 49–52.
———. "Glenn A. Black, 1900–1964." *American Antiquity* 31 (1966): 402–405.
Kelly, A. R. "Archaeology In The National Park Service." *American Antiquity* 5 (1940): 274–82.
———. "A Preliminary Report on Archaeological Explorations at Macon, Georgia." *Bureau of American Ethnology Bulletin* 119 (1938).
Kelly, Lawrence C. "Anthropology and Anthropologists in the Indian New Deal." *Journal of the History of the Behavioral Sciences* 16 (1980): 6–24.
Knowles, Nathaniel. "Cultural Stratification on the Trenton Bluff." *American Anthropologist* 43 (1941): 610–16.
Kroeber, A. L. "History And Science In Anthropology." *American Anthropologist* 37 (1935): 539–69.
———. "The Place of Anthropology in Universities." *American Anthropologist* 56 (1954): 464–67.
Kuper, Adam. "The Development of Lewis Henry Morgan's Evolutionism." *Journal of the History of the Behavioral Sciences* 21 (1985): 3–22.
Laudau, Larry. "The Pseudo-Science of Science." *Philosophy of the Social Sciences* 2 (1981): 173–98.
Leuchtenburg, William E. "The Pertinence of Political History: Reflections on the Significance of the State in America." *Journal of American History* 73 (1986): 585–600.

Lewis, T. M. N., and Madeline Kneberg. "The Prehistory of the Chickamauga Basin in Tennessee." *Tennessee Anthropology Papers* (1941).
Lister, Robert H. "Twenty Five Years of Archaeology in the Greater Southwest." *American Antiquity* 27 (1961): 39–45.
McKern, William C. "The Midwestern Taxonomic Method as an Aid to Archaeological Culture Study." *American Antiquity* 4 (1939): 301–13.
McNeil, William H. "Mythistory, or Truth, Myth, History, and Historians." *American Historical Review* 91 (1986): 1–10.
Meltzer, David J. "North American Archaeology and Archaeologists, 1879–1934." *American Antiquity* 50 (1985): 249–60.
———. "Paradigms and the Nature of Change in American Antiquity." *American Antiquity* 44 (1979): 644–57.
Mohrman, Harold. "Memoir of an Avocational Archaeologist." *American Antiquity* 50 (1985): 237–40.
Moore, James R. "Sources of New Deal Economic Policy: The International Dimensions." *Journal of American History* 61 (1974): 728–44.
Mulkay, Michael J. "Action and Belief on Scientific Discourse? A possible way of ending intellectual vassalage in social studies of Science." *Philosophy of the Social Sciences* 2 (1981): 163–71.
———. "Methodology in the sociology of science: Some reflections on the study of radio astronomy." *Social Science Information* 13 (1974): 107–19.
———. "The Mediating Role of the Scientific Elite." *Social Studies of Science* 6 (1976): 445–70.
Nash, Gerald D. "Herbert Hoover and the Origins of the Reconstruction Finance Corporation." *Mississippi Valley Historical Review* 46 (1959): 455–68.
Parsons, Talcott. "The Professions and Social Structure." *Social Forces* 17 (1939): 457–67.
Patterson, T. "The Last Sixty Years: Toward a Social History of Americanist Archaeology in the United States." *American Anthropologist* 88 (1986): 7–26.
Rae, John B. "Science and Engineering in the History of America." *Technology and Culture* 2 (1961): 391–99.
Reingold, Nathan. "Alexander Dallas Bache: Science and Technology in the American Idiom." *Technology and Culture* 11 (1970): 321–322.
Ritchie, William A. "Fifty Years of Archaeology in the Northeastern United States: A Retrospect." *American Antiquity* 50 (1985): 412–20.
Rossi, Paolo. "Baconianism." In *Dictionary of the History of Ideas: Studies of Selected Pivotal Ideas* 1 (1973): 172–79.
Sabloff, Jeremy A. "American Antiquity's First Fifty Years: An Introductory Comment." *American Antiquity* 50 (1985): 228–36.
Salmon, Merrilee H. "Ascribing Functions to Archaeological Objects." *Philosophy of the Social Sciences* 2 (1981): 19–26.
———. "'Inductive' versus 'Deductive' Archaeology." *American Antiquity* 41 (1976): 376–81.
Setzler, Frank M. "Archaeological Explorations in the United States: 1930–1942." *Acta Americana* 1 (1943): 206–20.
———. Introduction to "Archaeology of the Funeral Mound Ocmulgee National Monument, Georgia." *Archaeological Research Series, No. 3, National Park Service* 3 (1956): 1.
Setzler, Frank M., and Jesse D. Jennings. "Peachtree Mound and Village Site Cherokee County, North Carolina." *Bureau of American Ethnology Bulletin* 131 (1941).

Setzler, Frank M., and W. Duncan Strong. "Archaeology and Relief." *American Antiquity* 1 (1936): 301–309.
Sheldon, A. E. "Foreword," and "Introduction." *Nebraska History Magazine* 21 (1940): 143–52.
———. "Introduction." *Nebraska History Magazine* 22 (1941): 159–61.
Snow, Charles E. "Condylo-Diasphysical Angles of Indian Humeri from North Alabama." *Geological Survey of Alabama* (1940).
———. "Two Prehistoric Indian Dwarf Skeletons from Moundville." *Geological Survey of Alabama* (1943).
Spaulding, Albert C. "Fifty Years of Theory." *American Antiquity* 50 (1985): 301–308.
Stephenson, Robert T. "Administrative Problems Of The River Basin Surveys." *American Antiquity* 28 (1963): 277–85.
Steward, Julian, and Frank M. Setzler. "Function and Configuration in Archaeology." *American Antiquity* 28 (1963): 277–85.
Stocking, George W., Jr. "Editorial: on the Limits of 'Presentism' and 'Historicism' in the Historiography of the Behavioral Sciences." *Journal of the History of the Behavioral Sciences* 3 (1965): 211–17.
———. "From Physics to Ethnology: Franz Boas' Arctic Expedition as a Problem in the Historiography of the Behavioral Sciences." *Journal of the History of the Behavioral Sciences* 1 (1965): 53–66.
———. "The History of Anthropology: Where, Whence, Whither?" *Journal of the History of the Behavioral Sciences* 2 (1966): 281–90.
———. "The Santa Fe Style in American Anthropology: Regional Interest, Academic Institution, and Philanthropic Policy in the First Two Decades of the Laboratory of Anthropology, Inc." *Journal of the History of the Behavioral Sciences* 18 (1982): 3–19.
Strong, W. D. "Recommendations of the Committee on Basic Needs in American Archaeology." *Science* 90 (1939): 528–30.
Strong, W. D.; Waldo Wedel, ; J. L. Sellers; C. C. Osborne; and Marvin F. Kivett. "Asa T. Hill." *Nebraska History Magazine* 34 (1953): 67–90.
Thoresen, Timothy H. H. "Art, Evolution, and History: A Case Study of Paradigm Change in Anthropology." *Journal of the History of the Behavioral Sciences* 13 (1977): 107–25.
Turner, Stephen P. "Interpretive Charity, Durkheim, and the 'Strong Programme' in the Sociology of Science." *Philosophy of the Social Sciences* 2 (1981): 231–43.
Voget, Fred W. "Progress, Science, History, and Evolution in Eighteenth and Nineteenth-Century Anthropology." *Journal of the History of the Behavioral Sciences* 3 (1967): 132–55.
Webb, William S. "An Archaeological Survey of the Norris Basin in Eastern Tennessee." *Bureau of American Ethnology Bulletin* 118 (1938).
———. "An Archaeological Survey of Wheeler Basin on the Tennessee River in Northern Alabama." *Bureau of American Ethnology Bulletin* 122 (1939).
———. "The C. and O. Mounds at Paintsville, Sites Jo2 and Jo9, Johnson County, Kentucky." *The University of Kentucky Reports in Anthropology and Archaeology* 5 (1942).
———. "The Carlson Annis Mound, Site 5, Butler County, Kentucky." *The University of Kentucky Reports in Anthropology and Archaeology* 7 (1950).
———. "Indian Knoll, Site Oh2, Ohio County, Kentucky." *The University of Kentucky Reports in Anthropology and Archaeology* 4 (1946).

———. "The Jonathan Creek Village Site 4, Marshall County, Kentucky." *The University of Kentucky Reports in Anthropology and Archaeology* 8 (1952).
———. "Lexman Mound, Site Be32." *The University of Kentucky Reports in Anthropology and Archaeology* 5 (1943).
———. "The Morgan Stone Mound, Site 15, Bath County, Kentucky." *The University of Kentucky Reports in Anthropology and Archaeology* 5 (1941).
———. "Mt. Horeb Earthworks, Site 1, and the Drake Mound, Site 11, Fayette County, Kentucky." *The University of Kentucky Reports in Anthropology and Archaeology* 5 (1941).
———. "The Parrish Village Site, Site 45, Hopkins County, Kentucky." *The University of Kentucky Reports in Anthropology and Archaeology* 7 (1951).
———. "The Read Shell Midden, Site 10, Butler County, Kentucky." *The University of Kentucky Reports in Anthropology and Archaeology* 7 (1950).
———. "The Riley Mound, Site Be15, and the Landing Mound, Site Be17." *The University fo Kentucky Reports in Anthropology and Archaeolog* 5 (1943).
———. "The Wright Mounds, Sites 6 and 7, Montgomery County, Kentucky." *The University of Kentucky Reports in Anthropology and Archaeology* 4 (1940).
Webb, William S., and David L. DeJarnette. "An Archaeological Survey of Pickwick Basin in the Adjacent Portions of the States of Alabama, Mississippi and Tennessee." *Bureau of American Ethnology Bulletin* 129 (1942).
———. "Little Bear Creek Site Ct°8, Colbert County, Alabama." *Geological Survey of Alabama* (1948).
———. "The Perry Site Lu°25, Units 3 and 4, Lauderdale County, Alabama." *Geological Survey of Alabama* (1948).
Webb, W. S., and J. B. Elliott. "The Robbins Mounds, Sites Be3 and Be14, Boone County, Kentucky." *The University of Kentucky Reports in Anthropology and Archaeology* 5 (1942).
Webb, W. S., and W. D. Funkhouser. "The Chilton Site In Henry County, Kentucky." *The University of Kentucky Reports in Anthropology and Archaeology* 3 (1937).
———. "The Ricketts Revisited, Site 3." *The University of Kentucky Reports in Anthropology and Archaeology* 4 (1940).
———. "Rock Shelters in Menifee County, Kentucky." *The University of Kentucky Reports in Anthropology and Archaeology* 3 (1936).
Webb, W. S., and W. H. Haag. "Archaic Sites in McLean County, Kentucky." *The University of Kentucky Reports in Anthropology and Archaeology* 7 (1947).
———. "The Chiggersville Site, Site 1, Ohio County, Kentucky." *The University of Kentucky Reports in Anthropology and Archaeology* 4 (1939).
———. "Cypress Creek Villages, Sites 11 and 12." *The University of Kentucky Reports in Anthropology and Archaeology* 4 (1940).
———. "The Fisher Site, Fayette County, Kentucky." *The University of Kentucky Reports in Anthropology and Archaeology* 7 (1947).
Wedel, Waldo R. "Archaeological Investigations At Buena Vista Lake, Kern County, California." *Bureau of American Ethnology Bulletin* 130 (1941).
———. "An Introduction to Pawnee Archaeology." *Bureau of American Ethnology Bulletin* 112 (1936).
Wedel, Waldo R., and A. T. Hill. "Excavations at the Leary Indian Village and Burial Site, Richardson County, Nebraska." *Nebraska History Magazine* 17 (1936): 3–6.
Whately, Warren C. "Labor for the Picking: The New Deal in the South." *Journal of Economic History* 43 (1983): 905–29.

Whitely, Richard. "Umbrella and Polytheistic Scientific Disciplines." *Social Studies of Science* 6 (1976): 471–97.
Whittaker, Elvi. "Anthropological Ethics, Fieldwork and Epistemological Disjunctures." *Philosophy of the Social Sciences* 2 (1981): 437–51.
Willey, Gordon R. "Archaeology on the Florida Gulf Coast." *Smithsonian Miscellaneous Collections* 113 (1949).
———. "James Alfred Ford, 1911–1968." *American Antiquity* 24 (1969): 63–64.
———. "Some Continuing Problems in New World Culture History." *American Antiquity* 50 (1985): 351–63.
Woodbury, Richard B. "Regional Archaeological Conferences." *American Antiquity* 50 (1985): 434–44.

Unpublished Works

Lyon, Edwin Austin, II. "New Deal Archaeology In The Southeast: WPA, TVA, NPS, 1934–1942." Ph.D. dissertation, Louisiana State University, 1982.
Marlowe, Gregory Jon. "Preservation and Profession in American Archaeology: From 'Basic Needs' to the Committee for the Recovery of Archaeological Remains, 1939–1945." Paper presented at the Pacific Coast Branch, American Historical Association, Annual Meeting, San Diego, California, 11 August 1983.
Meltzer, David J. "Anthropology In Transition: 19th Century Foundations and Early 20th Century Changes." Paper Presented at the Pacific Coast Branch, American Historical Association, Annual Meeting, San Diego, California, 11 August, 1983.
Willey, Gordon R. "The Social Uses of Archaeology." The Kenneth B. Murdock Lecture. Leverett House, Harvard University, March, 1980.

Index

Abbott, Charles, 9-10, 94
Adams, Robert, 67
Adams, Robert McCormick Adams, 156
Addock, Julie, 113
Adena culture, 104, 105
Alabama, 111-13, 114-16; Baldwin, 116; Clarke County, 116; Colbert Creek site, 116; Guntersville Basin, 112, 116; Madison, 116; Mobile, 116; Pickwick Basin, 112, 115, 116; Wheeler Basin, 112, 113; Wheeler Dam, 67; Whitesburg Bridge, 116; Wright village site, 116
Alabama Anthropological Society, 112
Alabama Museum of Natural History, 14, 67, 112
Algonkian culture, 147
Algonquin culture, 133
American Anthropological Association (AAA), XXIV, 7, 44
American Antiquarian, XXV
American Antiquity, 63, 158
American Anthropologist, 6, 61
American Association for the Advancement of Science, XXV, 4, 62
American Council of Learned Societies, 1
American Ethnological Society, XXIV
American Indians, 9-10, 15. *See also individual Indian groups*
American Museum of Natural History, XXIV
American Naturalist, XXV
Anderson, Harold, 113

Anthropological Society of Washington, XXV
Antiquities Act of 1906, XXVII, 8
Archaeology: golden age of, 19; popularization of, 30, 37, 55-56, 158-59; salvage, 51; as science, XII, XIII, XVIII
Archaic culture, 104, 108, 116
Arizona, 140-41
Arizona Anthropological Association, 141
Arkansas, 158
Army Corps of Engineers, 21, 126
Atomic Energy Commission, XI

Baconian method (Francis Bacon), XVII, XXI, XXII
Barnes, George D., 51
Bartlett, Katherine, 113
Bell, Earl H., 135, 145
Berman, Joseph, 146
Berry, Brewton, 134
Birmingham Southern College, 112
Birmingham Anthropological Society, 112
Black, Glenn A., 148
Blue Eagle, 100
Boas, Franz, XXIII-XXIV, XX, XXV
Bohannan, C.T.R., 104
Bowden, A.O., 142
Bravo Valley Aspect, 152
Brew, J.O., 124
Brown, Ralph D., 104
Brown v. Board of Education, XI

220 / INDEX

Buckner, J.L., 104
Bulletin of the Texas Archaeological and Paleontological Society, 152
Bureau of American Ethnology (BAE), XXI, XXV, XXVII, 16, 48
Byers, Douglas, 147

C&O Mounds, 105
Caldwell, Joseph, 118
California: Buena Vista Lake site, 28-30; CWA sites in, 28; first projects, 26; Irvine Ranch, 150; Mohave, 150; Sacramento valley, 150; San Diego mountains, 150; San Joaquin valley, 150; Tehachapi mountains, 150; upper Newport Bay district, 150; Yokuts village, 31
Calusa Indians, 38
Carnegie Foundation, 6, 150
Carnegie Institution, 135, 151
Centralization of archaeology: of funding and relief programs, 97; lack of under WPA, 83-84; National Park Service in Georgia and, 118; New Deal contribution to, 60-61; relief policy and, 84
Chambers, Moreau B., 44, 134, 157
Champe, John L., 146
Cherokee culture, 46, 65
Chickasaw Indians, 134
Choctaw Indians, 134
Christian, Wayne G., 153
Civil Works Administration (CWA), 19-57; criticisms of, 85; demise of, 83; established precedent, 54; in Florida, 33; in Georgia, 117; publicity, 55; sets precedent for relief programs, 56; in Tennessee, 47
Civilian Conservation Corps (CCC): Arizona projects, 141; Kelly as supervisor, 118-19; in Kentucky, 106; in Mississippi, 157; National Park Service and, 17; in Nevada, 135; in New Mexico, 142; in North Dakota, 156; in Northwest, 61, 150; proposed survey of Central Mississippi Valley, 127

Classification and taxonomic systems: centralization of discipline and, 60; emphasis later criticized, 129-30; Gladwin, 136-37; in Louisiana, 102; Midwestern Taxonomic Method, 8, 67; models result from, 5; Plains Archaeological conferences and, 145; from TVA experience, 64; use of ceramics for, 66; of Wheeler Dam materials, 67, 68
Clearing House for Southwest Museums, 141
Clements, Forrest E., 153
Coe, Joffre, 28, 158
Colburn, Burnham, 49, 51
Colburn, W.D., 28
Colburn, William B., 32
Cole Creek complex, 158
Cole, Fay-Cooper: career, 11-12; Committee for Basic Needs, 124; organizes Ceramic Repository, 64-65; tree-ring dating, 67, 148; Webb's leadership and, 106
Coles Creek Horizon, 102
Collins, Henry, 25, 41
Colton, Harold S., 140
Committee on Accurate Publicity for Anthropology, 6-7
Committee for Basic Needs (CBN) on American Archaeology, 124-26
Committee for the Recovery of Archaeological Remains (CRAR), 126
Committee on State Archaeological Surveys (CSAS), 6, 9
Connecticut, 155
Cooper, Paul, 146
Copper-Galena complex, 67, 115-16
Cotter, John, 104
Cox, P.E., 45
Cressman, L.S., 150-51
Curruque Indians, 39

Dahms, Harold H., 113
Darwinism, XXI, XXIII
David, Robert B., 149
Davis, Elliot, 67

Day, H. Summerfield, 113
De Soto, Hernando, explorations by, 15, 32, 42
Deignan, Stella L., 108, 114, 115, 125
DeJarnette, David, 112-13, 114, 116
Delaware, 155
Dellinger, S.C., 158
Demeray, A.E., 89
Deuel, Thorne, 148
Disher, Kenneth B., 67
Dixon, Roland B., 6
Doran, Edwin B., Jr., 100
Douglass, A.E., 137
Draper, E.S., 48, 49
Drucker, Philip, 151
Dunlevy, Marion L., 113

Eisley, Loren C., 157
Elliott, John B., 104

Fairbanks, Charles, 41
Federal Civil Works Administration, 21-22
Federal Emergency Relief Administration (FERA), 59-69; CWA as alternative to, 20; demise of, 83; funds Marksville digs, 25; Hopkins chooses personnel from, 21; in Iowa, 157; in Mississippi, 157; in Missouri, 156; National Park Service proposal and, 17; in New Mexico, 134, 142; significance of projects, 133; in Tennessee, 47; Webb applies for labor to, 51
Fewkes, Jesse, XXV
Fewkes, Vladimir J., 118, 119, 120, 146-47
Field Museum (Chicago), XXIV, 151
Fieldwork: amateur, XXV-XXVI, 62; Chicago field method, 11-12, 41; early emphasis on, XXIII; first relief experiment, 23-26; Guthe on technique, 24; methods questioned, 68-69; models of New Deal archaeology, 23-26; as part of process, 52; Webb approach to, 50-51

Finney, Ron, 17
Florida: Belle Glade site, 37-38; Canaveral Island, 35, 39; Cocoa site, 35; competition for labor in, 35; CWA sites in, 28; first projects, 26; Hillsborough county, 158; Little Manatee River site, 38, 39; Palma Sola site, 38; Perioco Island, 38; Stirling directs digs, 33; Volusia County site, 39; West Florida, 158; West Palm Beach site, 34
Ford, James A.: assists Kelly, 26, 28, 40-42; background, 25, 100; career path, 97; uses Chicago field method, 41; contributions of, 99; Copena complex, 67; at Louisiana State University, 101; managerial paradigm and, 111; on popularization of archaeology, 56; proposed survey of Central Mississippi Valley, 126-27; on role of ceramics, 41; training at University of Michigan, 9; Webb's leadership and, 106
Fort Ancient culture, 46, 103, 104, 105, 155
Forum, XXV
Foster, James R., 113
Functionalism, conjunctive, 129-30
Funkhouser, William D., 50, 65, 68, 133

Gale, Bennett T., 67
Gayton, Anna H., 113
Geology: in Alabama, 14; John Wesley Powell and, XXI; stratigraphic excavation and, XXIV; systematizing archaeology and, XXIII; Walter Jones as state geologist, 14
Georgia, 116-21; Chatham county, 120; CWA sites in, 28; first projects, 26; Glynn county, 120; Irene Mound, 120; Macon sites, 12-17, 40-44, 55, 117, 120, 133-34; Marksville site, 24; Ocmulgee, 120; publicity, 116-17; St. Simons Island, 120
Georgia State Civil Works, 44
Gladwin, Harold S., 136, 140

222 / INDEX

Gladwin, Winifred, 136
Goodman, George H., 91-92
Goslin, Robert, 51
Government Accounting Office, 88
Great Depression, initial impact on research, 12
Greenacre, James C., 104
Griffin, James B.: on Adena culture, 105; analyzes field guide, 115; career path, 97; classification system, 66, 129; Copena complex, 67; on Ford using Chicago field method, 41; group for cooperative endeavors, 126; managerial paradigm and, 111; in Nebraska, 146; Norris Basin ceramics, 65-66; synthesis of Fort Ancient culture, 103; trained at University of Michigan, 9; Webb's leadership and, 106; Wheeler Dam shards examined, 68
Guthe, Carl: background, 9-11; on basic needs of archaeology, 123-24; contribution to organization of, 9; founding of Society for American Archaeology, 61-63; guest instructor in Birmingham, 112; leadership of state surveys, 6, 113; notes synthesis, 130-31; organizes Ceramic Repository, 64-65, 158; T.M.N. Lewis and, 108; on TVA project, 48, 49

Haag, William G.: on carbon-14 method, 130; at Louisiana State University, 101; managerial paradigm and, 111; opportunities as result of WPA, 97, 99; working with Webb, 51, 104
Hale, George Ellery, 5
Halseth, Odd S., 140
Harrington, M.R., 134
Harris, Walter, 14, 15, 44
Harrison, Ross, 124
Harrold, Charles: assists with credit, 44; expansion of activities, 43; as founding member of Society for Georgia Archaeology, 14, 15, 17; secures access to sites, 27

Haury, Emil: at Gila Pueblo, 135-38, 140; career, 137; classification systems, 129-30; FERA project experience, 133
Hawley, Florence, 67, 113, 148
Henry, A.V., 43
Henry, Joseph, XIX, XX, XXI, XXII, 36
Hewett, Edgar, 142-43
Hickock, Lorena, 86
Hill, A.T., 135, 144, 145
Hinsley, Curtis, XVII
Historic Sites Act of 1906, 92
Historic Sites Survey, 94
Hohansen, Theodore L., 113
Hohokam culture, 141
Holder, Preston, 100, 120
Holmes, William H., XX
Hopewell culture, 105
Hopewell-Woodland culture, 25
Hopkins, Harry: approves first projects, 26; background, 21; CWA criticized, 86-87; oversees Civil Works Administration, 21-23; types of WPA projects, 85
Horton, Donald, 146
Hoskins, James D., 107
Hrdlicka, Alex, XX

Ickes, Harold, 94
Illinois, 148
Illinois State Museum, 148
Indiana, Angel Mounds, 148
Iowa, 157

Jackson, A.T., 152
Jemez Field School, 142
Jennings, Jesse: assists Arthur Kelly, 41; career path, 97; criticized, 109; managerial paradigm and, 111; report on North Carolina by, 32; Smithsonian supervisor in North Carolina, 28, 32; at Volusia County site, 39
Johnson, Frederick, 147
Johnston, C. Stuart, 135, 152-53
Johnston, Claude, 104

Jones, Walter: Arthur Kelly and, 114-15; at Alabama Museum of Natural History, 112; career, 14; Georgia sites, 43; at Wheeler Basin (TVA), 49, 67
Judd, Neil: on amateur digs, XXVI; at TVA excavations, 49; on Committee for Accurate Publicity for Anthropology, 7; on World War, I, XXV

Kelly, A.R.: directs Archaeological Sites Division, 93; directs Macon sites, 133-34
Kelly, Arthur: Alabama expeditions, 114-15; character of, 118-20; Georgia projects, 26, 28, 40-41, 55, 117; Historic Sites committee, 94; at University of Georgia, 121
Kelly, J.C., 152
Kentucky, 103-7, 133
Kerr, Florence, 55, 124
Kidder, Alfred V.: at Carnegie Institution, 11; classification system, 66, 129; Committee for Basic Needs, 124; founding of Society for American Archaeology, 63; Gladwins meet, 136; leadership of state surveys, 6; pioneers stratigraphy, XXIV; on World War, I, XXV
King, Arden R., 100
Kivett, Marvin, 145
Kraxberger, Wayne W., 113
Krieger, Alex, 151
Krieger, Herbert W., 135
Kroeber, A.L., 6, 63
Kuhn, Thomas, XVIII

de Laguna, Frederica, 113
Lamar, Mrs. Walter D., 43
Lasseter, Roy, 140
Leadership, XV
Lee, Ronald, 94, 143
Lewis, Thomas M.N., 51, 107-11
Lippincott's Magazine, XXV
Lockard, D.W., 67
Louisiana: Crooks Mound site, 102; Lake Pontchartrain, 133; LaSalle-Paris site, 102; Louisiana State University as training center, 99-100, 101; significance of sites, 99-100; Troyville, 101

McGhee, W.H., XX
McIntyre, Lucy, 120
McKern, William, 8, 49, 124, 148
Maine, 155
Mallory, Jim, 14
Managerial paradigm: of archaeologic research process, 2; Ford and, 111; in Nebraska, 144; rationale for, XIII; universities and, XXIII
Mandan Indians, 156
Manhattan Project, XI
Marksville culture, 102
Marksville model, 23
Mason, Otis T., XX
Meleen, Elmer E., 156
Midwest, 143-146
Midwestern Taxonomic Method, 8, 106, 130
Miller & Lux, 29
Miller, Arthur, 50
Miller, Carl F., 104, 113
Miller, Raymond, 153
Miner, Horace, 67
Minnesota, 156-57
Mississippi, 134, 157
Mississippian culture, 32, 104, 106, 108, 115-16, 156
Missouri, 134, 155-56
Missouri Archaeological Society, 156
Missouri River Basin Survey, 28
Mogollon culture, 141
Monongahela Woodland culture, 147
Montana, 148-49
Morgan, Harcourt, 49
Morgan Stone Mound, 105
Morrison, Robert D., 67
Mount Horeb Earthworks, 105
Mulloy, William T., Jr., 100, 148
Munger, Paul, 156

National Academy Act of 1863, 5
National Academy of Sciences, 5

National Industrial Recovery Act (NIRA), 21
National Museum: in 19th century, XVII, Anthropological Society of Washington and, XXV; artifacts from Peachtree site, 32; artifacts from Shiloh site, 45; artifacts from Yokuts village, 31; compared to academe, XXIII; founding, XIX-XX; in Kansas, 157; prestige of, 16
National Park Service: Archaeological Sites Division created, 93; Civil Works Administration and, 21; controversy over New Mexico, 143; employment policies, 2-3; in Georgia, 117; growing responsibilities of, 86, 88-89; historic preservation role of, 92-93; in Nevada, 135; in New Mexico, 142; in Northwest, 61; Office of the Chief Park Historian, 93; post-World War II salvage operations, 126; proposal to use CCC labor or FERA funds in national parks, 17; proposed survey of Central Mississippi Valley, 127; Pueblo Grande Laboratory, 141; Setzler on advisory board, 94; shift of Smithsonian Institute power to, 92; Shiloh site, 45; Society for Georgia Archaeology and, 44; in Utah, 135
National Recovery Administration, 100
National Research Council: Birmingham conference, 14; as communication vehicle, 1; functions of, 5-6; grant to University of Kentucky, 50; lack of role in relief archaeology, 44; organizational expansion of archaeology and, 7-8; P.E. Cox and, 45; state surveys published by, 61; Subcommittee on the Archaeology of the Tennessee Valley, 49
National Youth Authority (NYA), 61, 147, 150, 151, 156
Nationalism, XVIII, XX
Naval Consulting Board, 5
Nebraska, 135, 143-46, 159

Nebraska Historical Society, 144-46
Neitzel, Robert S., 100, 109
Neuman, Marshall T., 113
Nevada, 135
New Deal, employment, 53-54
New Deal, First, 20
New Deal, Second, 21, 59
New England, 147, 155
New Hampshire, 155
New Jersey, 154-55
New Mexico, 141-43; Abo, 142; Chaco Canyon, 134, 142; Gila Pueblo, 135-38; Jemez Field School, 134, 142; Kuaua, 142; Las Cruces, 142; Pecos, 142; Pueblo Grande, 142; Quarai, 142; Santa Fe, 134, 142
New Mexico Anthropologist, 142
New York, 155
Newell, Perry, 146
Newman, Marshall, 38-39
Norris Basin, 65, 65-67
Norris Dam, 49
Norris, Sen. George, 47
North Carolina: Caraway Creek, 158; CWA sites in, 28; first projects, 26; Frutchey Mound, 158; Frutchey village site, 158; Guasili, 32; Montgomery county, 158; Occaneechi village site, 158; Orange county, 158; Peachtree site, 31-32
North Dakota, 156, 159
North Dakota Historical Society, 156

Ocmulgee National Monument: digs at, 119; early plans for, 27, 40; legislation introduced for, 44; managed by Kelly, 119; Society for Georgia Archaeology and, 134
Ohio Hopewell culture, 105
Ohio Serpent Mound, 159
Okeechobee site. *See* Florida, Belle Glade site
Oklahoma, 134, 153-54
Oregon, 135, 150-51
Organization: 19th century, XVII-XXIII; Guthe's contribution to, 9; infrastructure of Works Progress

INDEX / 225

Administration, 87; of laboratory units, 101-2; quality control, 87-88; watershed, 131; World War I and, 5; of WPA sites, 89-94

Palateu Indians, 149
Pan Handle culture, 152
Pawnee Indians, 144, 145
Payayan culture, 141
Peabody Museum (Harvard University), XXIV, 151
Peachtree (North Carolina) site, 31-32
Pennsylvania, 146-47
Pennsylvania Archaeological Society, 17
Pennsylvania Historical Commission, 147
Petrullo, Vincenzo: Alabama expeditions, 114; background and career, 91-92; controversy over New Mexico projects, 142-43; Georgia expeditions, 117; standardization of methods, 94; T.M.N. Lewis and, 108
Phillips, Phillip, 126
Plains Archaeological conferences, 7, 145
Poffenberger, A.T., 10, 48
Pot-hunting, concern over, XXV-XXVI, 5, 63
Powell, John Wesley, XX-XXI, XXII, 3
Professionalization: academic paradigm, 3-4; collegial contact and, 98-99; conference format, 4-5; doctoral training and, XXIII; Lewis resists, 109-10; professional community grows, 122, 123; relief policy and, 84-85; standardized field practices during TVA work, 52
Public Works Administration (PWA), 22, 135
Pueblo culture, 141
Puerto Rico, 158

Quimby, George I., Jr., 9, 100

Read, William F., 153
Reagan, Albert B., 135

Reconstruction Finance Corporation, 21
Reed, Eric K., 39
Renger, J.J., 67
Rhode Island, 155
Ritchie, William, 104
Roberts, Frank H.H.: on breadth of field work, 141; in Colorado, 151; emphasis on classification, 129; founding of Society for American Archaeology, 63; on Smithsonian staff in Tennessee, 28, 44, 46
Roberts, Henry, 158
Rockefeller Foundation, 6
Roosevelt, Franklin D., 47, 83

Sabloff, Jeremy, 130, 136
Sayre, H.M., 148
Schwartz, Douglas, 105
Science, XXV 123-24, 124-25
Sellards, E.H., 153
Setzler, Frank M.: on advisory board of National Park Service, 94; first relief project at Marksville, 23, 24, 25-26; Historic Sites committee, 94; links to Georgia archaeologists, 15, 117-18; politics of CWA, 36, 37; procedural problems, 89; relationship with A.R. Kelly, 93; report on North Carolina by, 32; on Shiloh site, 46; Smithsonian supervision of sites, 28; visits Illinois, 148
Shepperson, Gay, 44, 55
Shoshoni culture, 149, 150
Smith, Carlyle, 100, 145
Smithsonian Institution: amateurs and, XIV; Anthropological Society of Washington and, XXV; as assimilationist, 36-37; autonomy challenged, 45; budget, XXI; as central body, 20; challenged by university-based archaeologists, 85; changing participation of, 61, 95-96; Civil Works Administration and, 21; as clearinghouse, 54; compared to academe, XXIII; CWA and, 36-37, 54; domination declines, 60, 90; domination of, XVII-XXII;

effects of Great Depression on, 12; on efficacy of relief projects, 57; employment policies, 2-3; excavation model, 23; first relief experiment, 22-23; first state projects, 26; founding of, XVIII-XIX; in Georgia, 14-15, 27, 117-18; importance of publication, 56, 64; incorporates Bureau of American Ethnology, XXI; in Kansas, 157; leadership, XXII; legislation sponsored in 1928, 9-10; National Park Service and, 17; in Nebraska, 144; in Northwest, 150; political strategies of, 20; post-World War II salvage operations, 126; power challenged, 48; prestige of, 16; role with WPA, 86; Society for American Archaeology and, XVI; supervision of sites by, 28; TVA publications, 64; use of public facilities, 17; WPA cooperation, 143
Snow, Charles E., 113
Society for American Archaeology: creation of Ocmulgee National Monument, 134; founded, 19, 61-63; founding of, 130-31; professional community and, 99; Smithsonian Institution and, XVI; T.M.N. Lewis and, 110; women in, 113
Society for Georgia Archaeology, 13-14; achievements of, 44; character of, 117; launching Macon digs, 16-17, 40; need for storage facilities, 43; talk on use of African-African women laborers, 120-21
Solomon, Lint, 14
South: depression conditions in, 22; relief programs, 53-54
South Dakota, 156, 159
Southeastern Archaeological conferences, 107
Sowers, Ted C., 149
Spaulding, Albert C., 9, 104, 156
Specialization, growth of, 64-65, 68
Spier, Leslie, 6-7

Standardization: academic archaeology and, 95; Chicago students facilitate, 12; Petrullo contributes to, 94; Petrullo and Deignan emphasized, 111; relief project development of, 25, 90-94; Smithsonian assists in, 54; T.M.N. Lewis and, 110; of training, 115; viewed as process, XIII-XIV; voluntary adherence to, 100
State Archaeologic Surveys: activities in 1934 and 1935, 135; artifacts from Spiro Mound sold, 154; Committee on State Archaeological Surveys (CSAS), 6-7; communication among colleagues by, 56; goal for expanded role of, 62; number of institutions included in, 8; Society for American Archaeology and, 63; Sub-committee on the Archaeology of the Tennessee Valley, 49
Stephens, Alden B., 67
Stirling, Gene, 28, 38, 52
Stirling, Matthew W.: at TVA excavations, 48-49; competition for labor at sites, 35; controversy over New Mexico, 143; on Cornfield Mound, 42; criticism by Van Hyning, 33, 36; first state projects, 23, 26; Historic Sites committee, 94; links to Georgia archaeologists, 15, 16-17; political problems of, 36-37, 53; Southeastern taxonomy by, 8; supervises Florida sites, 28, 32, 34, 35; training, 52; WPA standardization and, 94
Stout, David, 104
Stovall, J. Willis, 153
Strong, W. Duncan: at Columbia University, 147; chair of Committee for Basic Needs, 124; political problems of, 53; Smithsonian supervisor in California, 28, 29-30, 31, 120-21; ties to Nebraska, 135, 145
Studer, Floyd V., 152
Sullivan, H.M., 51
Sundance Museum, 149

Surruque Indians, 39
Swanton, John, 26, 31-32

Taylor, A.P., 51
Taylor, Walter W., 129-30
Tchefuncte culture, 102
Tennessee, 107-9; Chickamauga, 108-9; CWA sites in, 28; first projects, 26; Shiloh National Park site, 44-46; Tennessee River Valley site, 46-51, 63
Tennessee Valley Authority, 46-51; administrative aspects, 46-47, 68; archaeology model, 23-24; first phase of TVA archaeology, 44-45; publicity about sites, 99; theoretical and methodological breakthroughs, 60-61; Webb's leadership, 103
Terra, Helmut de, 151
Texas, 152-153
The Century Magazine, XXV
Training, standard procedures for, XIII-XIV
Treadway, C.B., 33

United States National Museum (USNM). *See* National Museum
University of Chicago field method, 11-12
University of Kentucky: archaeology established at, 103; significance of program, 106-7
University of Michigan: Ceramic Repository, 7-8, 64, 66, 67, 146; Museum of Anthropology, 9; unique contribution of, 158
University of Tennessee, 107
University of Texas Anthropological Papers, 152
Upper Republican culture, 145
U.S. Geological Survey (USGS), XXIV-XXV, 151
Utah, 135, 151

Van Hyning, T., 33, 36
Veterans Administration, 21
Vinson, Carl, 44

Walker, Wendell C., 51
Walker, Winslow, 28, 101, 148, 156
Warner Brothers Studios, 36
Washington (state), 151
Wauchope, Robert, 118, 121
Webb, William S.: campaigns to continue projects, 56; career and contributions, 49-51, 111; Committee for Basic Needs, 124; compared to Lewis approach, 108-9; contribution in Kentucky, 103-7; fundraising by, 104; hiring methods, 68; interpretative schemes, 105-6; leadership of, 66; oversees TVA labor, 63; specialized experts and, 64; supervises DeJarnette, 113, 116; taxonomic systems used by, 8, 129
Wedel, Waldo R.: Colorado expedition, 151; in Iowa, 157; Missouri sites, 156; Nebraska sites, 135, 144; Smithsonian supervisor in California, 28, 31
Wheat, Joe Ben, 152
Wheeler Dam, 49, 67
White, James W., 67
Wilder, Charles D., 51
Wilford, L.A., 156
Wilkie, A.E., 51
Willamette Valley expedition, 151
Willey, Gordon: assists Arthur Kelly, 41; assists Ford at LSU, 101; background, 100; career path, 97; Copena complex, 67; FERA project experience, 133; Florida sites, 158; on Georgia projects, 119; on Gladwin's classificatory system, 136; managerial paradigm and, 111; proposed survey of Central Mississippi Valley, 127; synthesizes depression-era digs, 40, 130; Webb's leadership and, 106; works with African-American labor, 120
Winterbourne, John W., 150
Wisconsin, 147-148
Wisconsin Archaeological Society, 147-48
Wisconsin State Historical Museum, 147-48
Wissler, Clark, 6, 124

Women: anthropologists, 10-11; black female labor on digs, 113, 114; in New Deal archaeology, 113-14; on Wyoming projects, 149

Women's and Professional Division (of Works Progress Administration): archaeology classified under by WPA, 54-55; enforcement of standards by, 100; Kerr initiative on decision-making, 124; Petrullo in, 91; reports to, 88, 92

Woodbury, George, 39

Woodbury, Richard, 158

Woodland culture, 32, 104, 108, 133, 156

Woodward, Ellen, 91-92, 94

Works Progress Administration, 83-96; in Arizona, 140; career opportunities and, 97; Committee for Basic Needs, 124; criticisms of relief and, 22; dismantled, 126; in Georgia, 117; laboratory funding, 119; in New Mexico, 141; Planning Committee, 125-26; publicity campaigns, 122, 144, 158-59; survey work in Nebraska, 135; in Tennessee, 47; transition to, 59; Women's and Professional Division (WPD), 54-55, 88, 91, 92, 100

Works Projects Administration. *See* Works Progress Administration

World War I, XXV

World War II, 114, 125

Wormington, H.M., 151

Wrench, Jesse, 134

Wyoming, 149-50

Yale Survey, 151

www.ingramcontent.com/pod-product-compliance
Lightning Source LLC
Chambersburg PA
CBHW030340240426
43661CB00052B/1695